Geographic information system (GIS) technology has the power to help *transform* our world. The maps in this year's *Esri Map Book* offer excellent examples of the GIS user community doing just that.

We at Esri applaud this work. The planet is facing difficult challenges—indeed, precarious, *geographic* challenges. The maps here demonstrate creative solutions to serious threats affecting our environment, our infrastructure, and our safety.

We need to keep pushing the boundaries of GIS to meet these challenges, and I'm confident in our collective abilities to do this work.

But how does that happen?

The "scaling up" of GIS—making it as pervasive as GPS—through web applications and mobile devices and the enrichment of this new framework for organizing, managing, integrating, and sharing data are all part of the effort. Critical to this paradigm, however, is the important and transformative work of GIS professionals.

When I look at a map, I am looking for a story. The stories in this book are inspiring, and I congratulate all contributors for their vision and commitment to inventing smarter realities and building a better world. GIS users undoubtedly will continue to develop innovative ideas for spatially based, geographic problem solving.

Warm regards,

Jack Dangermond

Table of Contents

World Purchasing Power

Michael Bauer Research GmbH
Nuremberg, Bavaria, Germany

By Michael Bauer Research GmbH

Contact

Sabine Petrasek, sabine.petrasek@mb-research.de

Software

ArcGIS 10.0 for Desktop

Data Sources

Michael Bauer Research GmbH, Eurostat, Statistics Estonia, Central Bureau of Statistics/Topographische Service, Emmen, INE, National Geospatial-Intelligence Agency, Bundesbehörden der Schweizerischen Eidgenossenschaft, US Census Bureau, Statistics South Africa, Statistics New Zealand, Commonwealth of Australia, Geography Division, Statistics Canada, Instituto Brasileiro de Geografia e Estatistica, contains Ordnance Survey data Crown copyright and database right

Every year, Michael Bauer Research GmbH creates a map that shows the large differences in per capita income worldwide on a national and subnational level. The map describes the disposable household income per capita (income without taxes and obligatory social security contributions, including received transfer payments) of a certain area's private households.

Purchasing power is not distributed homogeneously within each country but mostly concentrated in the economic centers. High per capita values are also located in resource-rich but often sparsely populated regions such as Russia's Siberia or the Australian Outback.

AE	United Arab Emirates		
AL	Albania		
AM	Armenia		
AT	Austria		
BA	Bosnia and Herzegovina		
BE	Belgium		
BH	Bahrain		
BI	Burundi		
CH	Switzerland		
DK	Denmark		
EE	Estonia	ME	Montenegro
HR	Croatia	MK	Macedonia
IL	Israel	MT	Malta
IT	Italy	NL	Netherlands
KW	Kuwait	QA	Qatar
LB	Lebanon	RS	Serbia
LT	Lithuania	SI	Slovenia
LU	Luxembourg	UK	United Kingdom
LV	Latvia	XK	Kosovo

Norway
Sweden
Finland
Oslo
Helsinki
Stockholm
Tallinn EE
DK
Copenhagen
Riga LV
NL
Berlin LT *Vilnius*
Germany Poland *Minsk*
Warsaw Belarus
Prague Czech Republic *Kyiv*
U *Vienna* Slovakia
CH AT *Budapest* Ukraine
HR *Bucharest* Moldova
Rome IT BA RS Romania
MC SI Bulgaria
AL *Sofia* *Istanbul*
Tunis MT Greece *Ankara* Georgia *Tbilisi*
MK *Nicosia* *Baku* AM Azerbaijan
Tunisia *Tripoli* Cyprus LB *Damascus* Syria Turkey
Jerusalem *Amman* Iraq
Cairo Jordan *Baghdad* Iran *Tehran*
Libya Egypt KW
BH QA
Riyadh AE *Abu Dhabi*
Saudi Arabia *Muscat*
Oman

Moscow

Russia

Astana

Kazakhstan

Uzbekistan *Almaty*
Bishkek Kyrgyzstan
Toshkent Tajikistan
Turkmenistan *Ashgabat* *Dushanbe*
Kabul
Afghanistan
Pakistan *Islamabad*
New Delhi Nepal
Kathmandu *Thimphu* Bhutan
India Bangladesh
Mumbai *Dhaka*
Myanmar

Mongolia
Ulaanbaatar

China

Beijing *Vladivostok*
North Korea
Pyongyang
Seoul South Korea Japan
Tokyo
Shanghai

Taipei
Taiwan
Hanoi *Hong Kong*
Laos
Naypyidaw
Thailand *Vientiane* Philippines
Bangkok Cambodia *Manila*
Phnom Penh Vietnam

Niger
Chad
N'djamena
Nigeria
Abuja
Sudan *Khartoum* Eritrea
Asmara *Sana'a*
Djibouti Yemen
Addis Ababa Somalia
Central African South Sudan *Juba* Ethiopia
Republic *Bangui*
Cameroon
Yaoundé
Guinea
Republic Uganda *Kampala* Kenya
Gabon of *Libreville* Rwanda *Nairobi*
Congo BI
Brazzaville Democratic
Kinshasa Republic
of the Congo Tanzania
Luanda *Dar es Salaam*

Maldives

Sri Lanka
Colombo

Mogadishu

Seychelles

Brunei *Bandar Seri Begawan*
Kuala Lumpur Malaysia
Singapore

Indonesia
Jakarta

Papua New Guinea
Port Moresby
East Timor *Dili*

Solomon Islands
Honiara

Angola
Zambia *Lusaka*
Malawi
Lilongwe
Zimbabwe Mozambique
Namibia Botswana *Harare*
Windhoek Gaborone
Maputo
Johannesburg Swaziland
South Africa Lesotho
Cape Town

Antananarivo
Madagascar
Mauritius
Réunion

Australia
Perth
Canberra
Sydney

New Zealand
Wellington

Vanuatu
Port-Vila
Nouméa
New Caledonia

Fiji
Suva

Apia
Samoa
Nuku'alofa Tonga

rchasing Power 2012
sposable Household Income)

o per capita and year
minal values)

no data
below 2,000
2,000 - 4,000
4,000 - 6,000

6,000 - 8,000
8,000 - 10,000
10,000 - 15,000
15,000 - 20,000

20,000 - 25,000
25,000 - 30,000
30,000 - 40,000
above 40,000

Scale at Equator
Kilometers
0   250 500   1.000   1.500   2.000

5

ROI Analysis

Mosaic
Irving, Texas, USA

By David Drury, Craig Massey, Andrew Green, Stephen Pope, and Joshua Chlapek

Contact

David Drury, david.drury@mosaic.com

Software

ArcGIS for Desktop, Esri Business Analyst

Data Sources

Mosaic, service layer ©2012 Esri, DeLorme, NAVTEQi

The Mosaic marketing agency conducts thousands of consumer events and retail visits across North America every year. To ensure a competitive advantage, Mosaic's mapping team leverages the Esri platform to get the right resource to the right place at the right time to most efficiently execute work, then provides visualization showing the results of the work.

In this analysis, multiple variables are used to model and prioritize locations with the highest probability of success for marketing activities. Marketing activities at the best locations can range from installing promotional materials in grocery stores to a nationwide concert series to publicize a new software release. Once the execution is completed, Mosaic's mapping team provides visualizations, such as weekly heat maps, to show the change in sales volumes after a promotional event.

Courtesy of Mosaic.

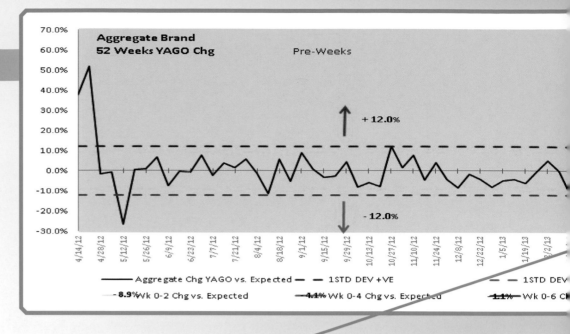

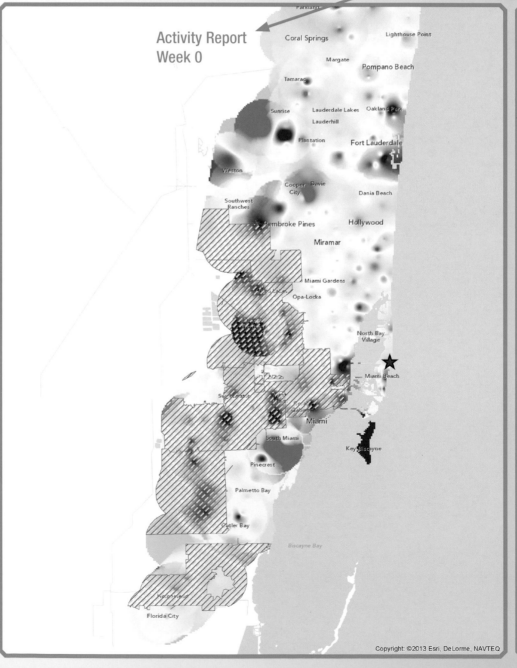

Activity Report
Week 0

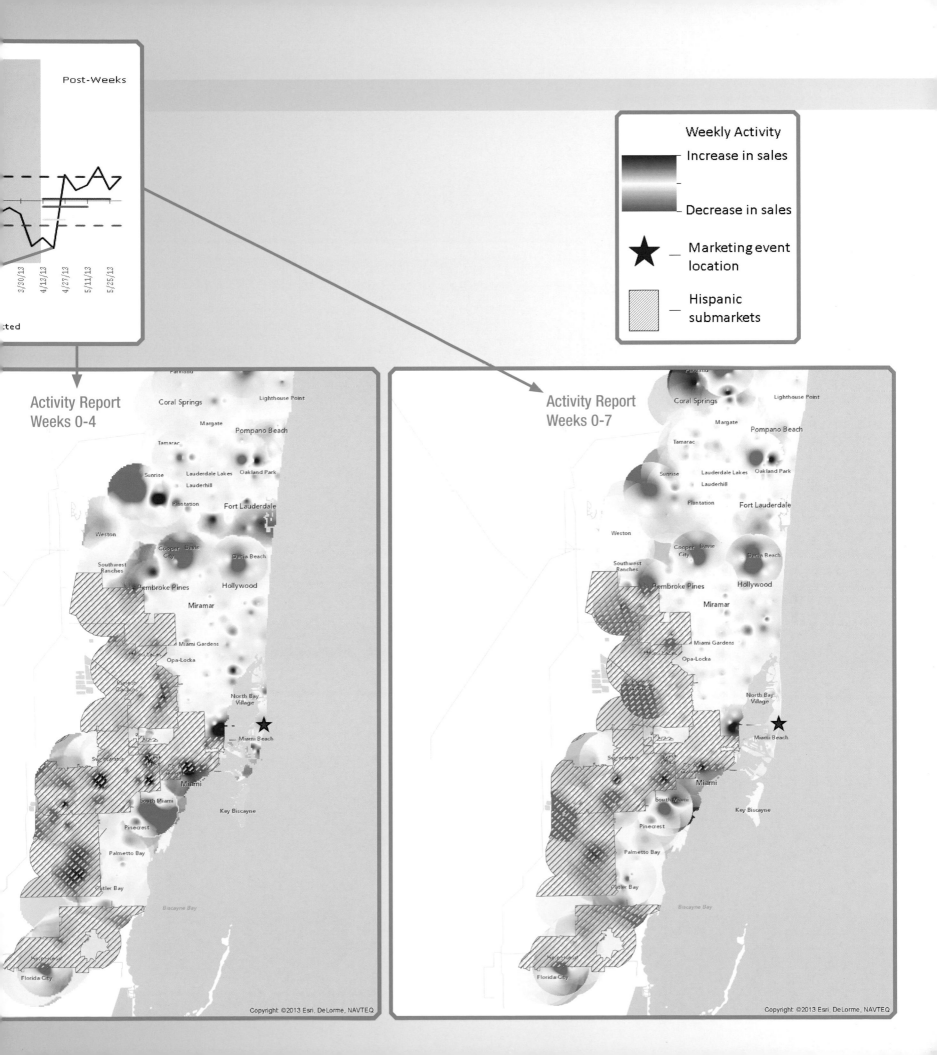

Post-Weeks

3/30/13 4/13/13 4/27/13 5/11/13 5/25/13

ted

Weekly Activity

Increase in sales

Decrease in sales

★ Marketing event location

Hispanic submarkets

Activity Report
Weeks 0-4

Parkland
Coral Springs   Lighthouse Point
              Margate
                        Pompano Beach
Tamarac
              Lauderdale Lakes Oakland Park
Sunrise   Lauderhill
         Plantation
                        Fort Lauderdale

Weston
      Cooper Davie
      City
Southwest   Dania Beach
Ranches
         Pembroke Pines
                        Hollywood
              Miramar

                  Miami Gardens
              Opa-Locka

                        North Bay
                        Village
                              ★
                        Miami Beach

Sweetwater
                  Miami
         South Miami
                  Key Biscayne

         Pinecrest

      Palmetto Bay

      Cutler Bay

              Biscayne Bay

Florida City
                    Copyright ©2013 Esri, DeLorme, NAVTEQ

Activity Report
Weeks 0-7

Parkland
Coral Springs   Lighthouse Point
              Margate
                        Pompano Beach
Tamarac
              Lauderdale Lakes Oakland Park
Sunrise   Lauderhill
         Plantation
                        Fort Lauderdale

Weston
      Cooper Davie
      City
Southwest   Dania Beach
Ranches
         Pembroke Pines
                        Hollywood
              Miramar

                  Miami Gardens
              Opa-Locka

                        North Bay
                        Village
                              ★
                        Miami Beach

Sweetwater
                  Miami
         South Miami
                  Key Biscayne

         Pinecrest

      Palmetto Bay

      Cutler Bay

              Biscayne Bay

Florida City
                    Copyright ©2013 Esri, DeLorme, NAVTEQ

Surficial Geology of Alberta

Alberta Energy Regulator—Alberta Geological Survey

Edmonton, Alberta, Canada

By Mark Fenton, Joan Waters, Steven Pawley, Nigel Atkinson, Daniel Utting, and Kirk Mckay

Contact

Joan Waters, joan.waters@aer.ca

Software

ArcGIS Desktop 9.3.1, InDesign CS6, GeoScaler, FME

Data Sources

Alberta Geological Survey field data, lidar data, aerial photography, topographic and previous geological paper copy maps at various scales; Fenton, M.M., Waters, E.J., Pawley, S.M., Atkinson, N., Utting, D.J., and Mckay, K. (2013): Surficial Geology of Alberta; Alberta Energy Regulator, AER/AGS Map 601, scale 1:1 000 000

This map portrays a generalized compilation of the surficial geology of Alberta (1:1,000,000 scale) using published Alberta Geological Survey, Geological Survey of Canada, and Environment Canada maps, as well as university theses and new data. Alberta Geological Survey staff members compiled vector data from sixty-five existing maps (scales 1:50,000 to 1:500,000) along with project-specific mapping of the few unmapped areas (1:500,000 to 1:1,000,000 scale). They incorporated all the data into a seamless provincial mosaic. Map boundary discrepancies were largely resolved during the process of reclassifying the input data to a common legend with only limited remapping. The staff members then generalized the mosaic using GeoScaler software published by the Geological Survey of Canada (Open File 6231). The resultant surficial geology layer was pan sharpened with a hillshaded Shuttle Radar Topography Mission digital elevation model to create the final map.

Courtesy of Alberta Energy Regulator—Alberta Geological Survey.

Alberta Geological Survey Map 601

Surficial Geology of Alberta

M.M. Fenton, E.J. Waters, S.M. Pawley, N. Atkinson, D.J. Utting and K. Mckay

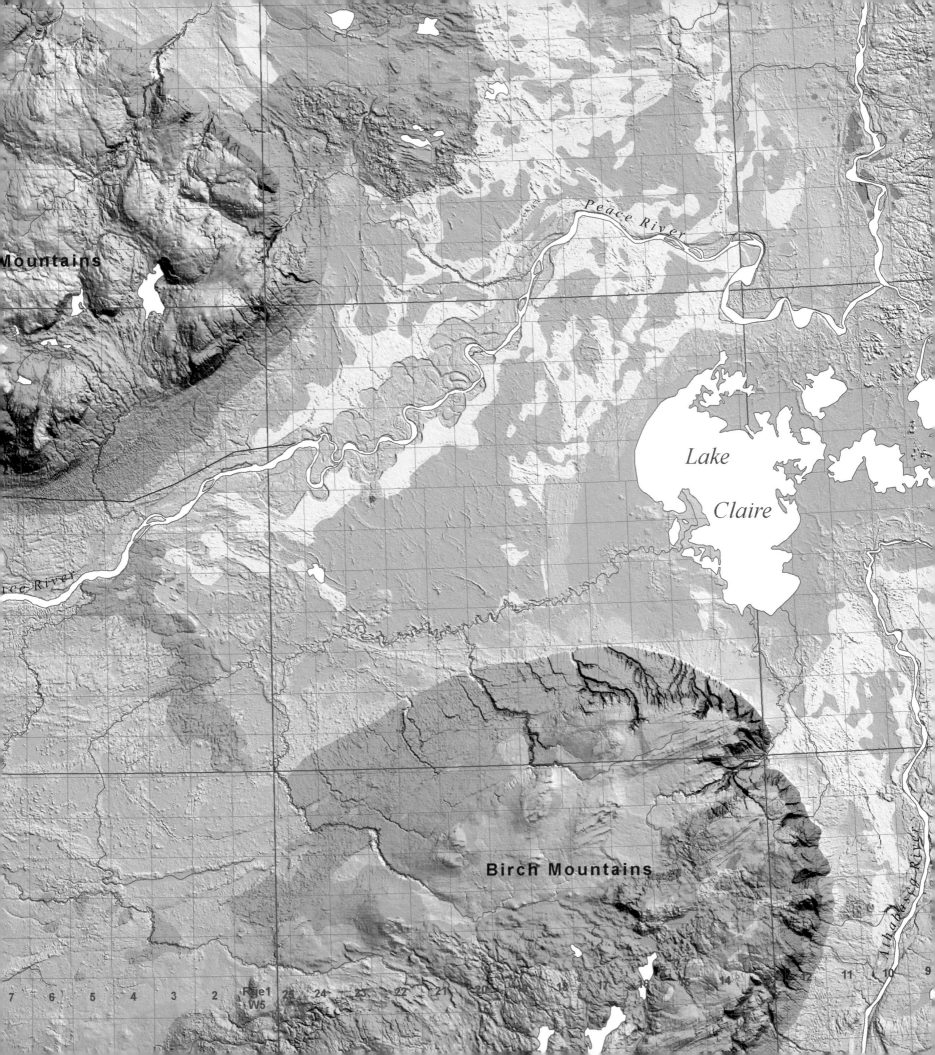

Mountains

Peace River

Lake

Claire

ce River

Birch Mountains

Athabasca River

7   6   5   4   3   2   nge1 26   24   23   22   21   20   18   17   14   12   11   10   9
                                   W5

2013 Map of Rio de Janeiro City

Instituto Pereira Passos—Prefeitura da Cidade do Rio de Janeiro (Pereira Passos Institute—Prefecture of Rio de Janeiro)

Rio de Janeiro, Brazil

By Instituto Pereira Passos/Gerência de Cartografia—Tematica Cartografia

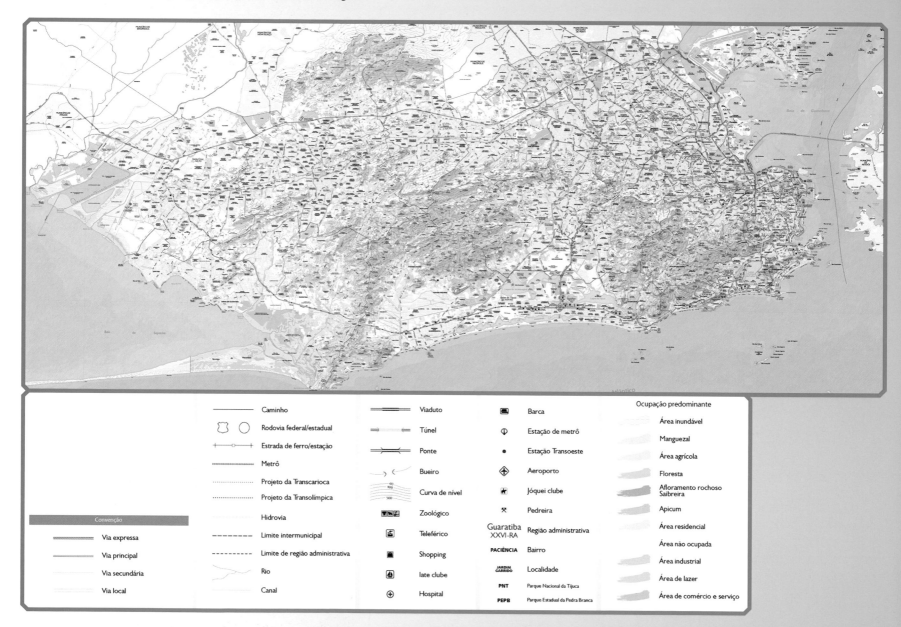

							Ocupação predominante
———	Caminho	≡≡≡	Viaduto	🔲	Barca		
⬡ ○	Rodovia federal/estadual	≡≡≡	Túnel	Φ	Estação de metrô		Área inundável
+—○—+	Estrada de ferro/estação	━━━	Ponte	•	Estação Transoeste		Manguezal
··········	Metrô	⌒ ⊂	Bueiro	◈	Aeroporto		Área agrícola
··········	Projeto da Transcarioca	〰	Curva de nível	⛊	Jóquei clube		Floresta
··········	Projeto da Transolímpica						Afloramento rochoso Saibreira
	Hidrovia	⛝	Zoológico	⚒	Pedreira		Apicum
-------	Limite intermunicipal	🏛	Teleférico	Guaratiba XXVI-RA	Região administrativa		Área residencial
-------	Limite de região administrativa	▣	Shopping	PACIÊNCIA	Bairro		Área não ocupada
〰	Rio	⚓	late clube	JARDIM GARRIDO	Localidade		Área industrial
——	Canal	⊕	Hospital	PNT	Parque Nacional da Tijuca		Área de lazer
				PEPB	Parque Estadual da Pedra Branca		Área de comércio e serviço

Convenção

≡≡≡	Via expressa
━━━	Via principal
───	Via secundária
───	Via local

Contact

Luiz Roberto Arueira da Silva, luiz.arueira@rio.rj.gov.br

Software

ArcGIS 10.0 for Desktop, Adobe Illustrator

Data Source

Instituto Pereira Passos—Prefeitura da Cidade do Rio de Janeiro

Pereira Passos Institute, the planning and information arm of Rio de Janeiro's municipal government, gathers, prepares, and disseminates data needed for planning and city management. This map of Rio de Janeiro City represents the main land uses of the city. It also contains information about the mass transportation lines, such as train and metro systems with their stations and Bus Rapid Transit systems in use and planned. The map also contains the toponyms of the whole city, with names of all the neighborhoods and historical locations. Rio de Janeiro has a population of 6.3 million, the second largest in Brazil, and is one of the most visited cities in the Southern Hemisphere. It will host the 2016 Summer Olympic Games.

Courtesy of Instituto Pereira Passos.

2013 Global Flight Network

BioDiaspora

Toronto, Ontario, Canada

By David Kossowsky

Contact

Dr. Kamran Khan, info@biodiaspora.com

Software

ArcGIS 10.1 for Desktop

Data Source

OAG 2013

Over the course of a year, there are more than 38,000 different flight routes that connect cities with one another. This map, created with Python and ArcGIS, provides a greater understanding of how interconnected the world is due to today's air transportation. Complex but often invisible networks link us together to create a truly globalized world. Made entirely out of flight lines, this image shows not only transportation links but also major airport hubs and highlights the differences between industrialized nations and the developing world.

Courtesy of BioDiaspora.

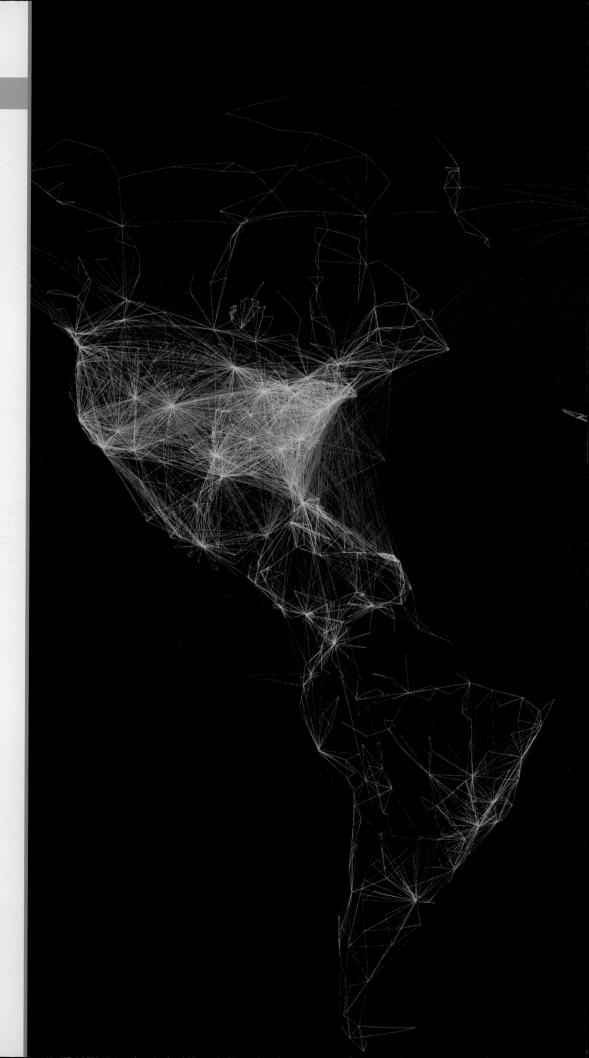

Duration of Travel

<1   (hours)   >20

Using Lidar-Derived DEMs for Precision Agricultural Conservation Planning

US Department of Agriculture (USDA) Agricultural Research Service, National Laboratory for Agriculture and the Environment
Ames, Iowa, USA

By David James, Sarah Porter, Mark Tomer, and Eileen McLellan

Contact

David James, david.james@ars.usda.gov

Software

ArcGIS 10.1 for Desktop

Data Source

Iowa Department of Natural Resources

Improving the quality of water discharged from agricultural watersheds requires comprehensive and adaptive approaches for planning and implementing conservation practices. These measures will need to consider landscape hydrology, distributions of soil types, land cover, and crop distributions in a comprehensive manner. The two most consistent challenges to these efforts will be consistency and reliability of data and the capacity to translate conservation planning from watershed to farm and field scales. The translation of scale is required because, while conservation practices can be planned based on a watershed scale framework, they must be implemented by landowners in specific fields and riparian sites that are under private ownership. The US Department of Agriculture's Agricultural Research Service National Laboratory for Agriculture and the Environment (NLAE) and Environmental Defense Fund have partnered to develop planning approaches, high-resolution spatial datasets, and conservation practice assessment tools that will allow the agricultural and conservation communities to characterize and mitigate these challenges. This multistate spatial database can be used as a planning framework to evaluate land use, landscape hydrology, and soil distributions.

In this example, the Beaver Creek, Iowa, watershed (HUC12) has been analyzed to showcase possible placement scenarios for two conservation practices, Nutrient-removal Wetlands and Resaturated Riparian Buffers. The HUC12-based conservation planning database incorporates high-resolution, lidar-derived digital elevation models (3 m); Soil Survey Geographic database soils (10 m); and the National Agricultural Statistics Service Cropland Data Layer (30 m) over a field boundary framework for more than 4,000 Iowa, Illinois, and Minnesota watersheds.

Courtesy of USDA ARS National Laboratory for Agriculture and the Environment.

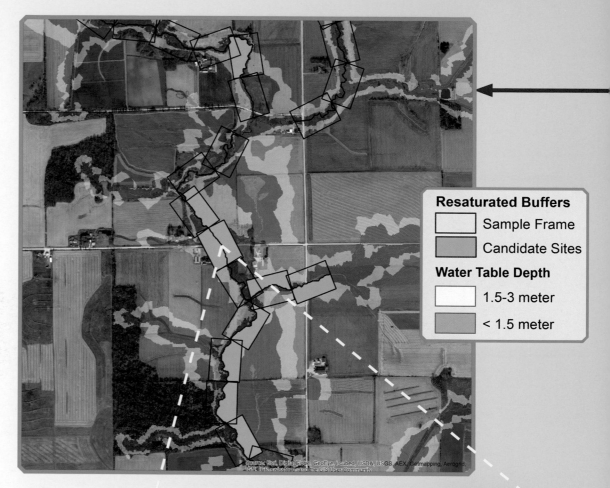

Source: Esri, DigitalGlobe, GeoEye, i-cubed, USDA, USGS, AEX, Getmapping, Aerogrid, IGN, IGP, swisstopo, and the GIS User Community

Resaturated Buffers

	Sample Frame
▓	Candidate Sites

Water Table Depth

	1.5-3 meter
▓	< 1.5 meter

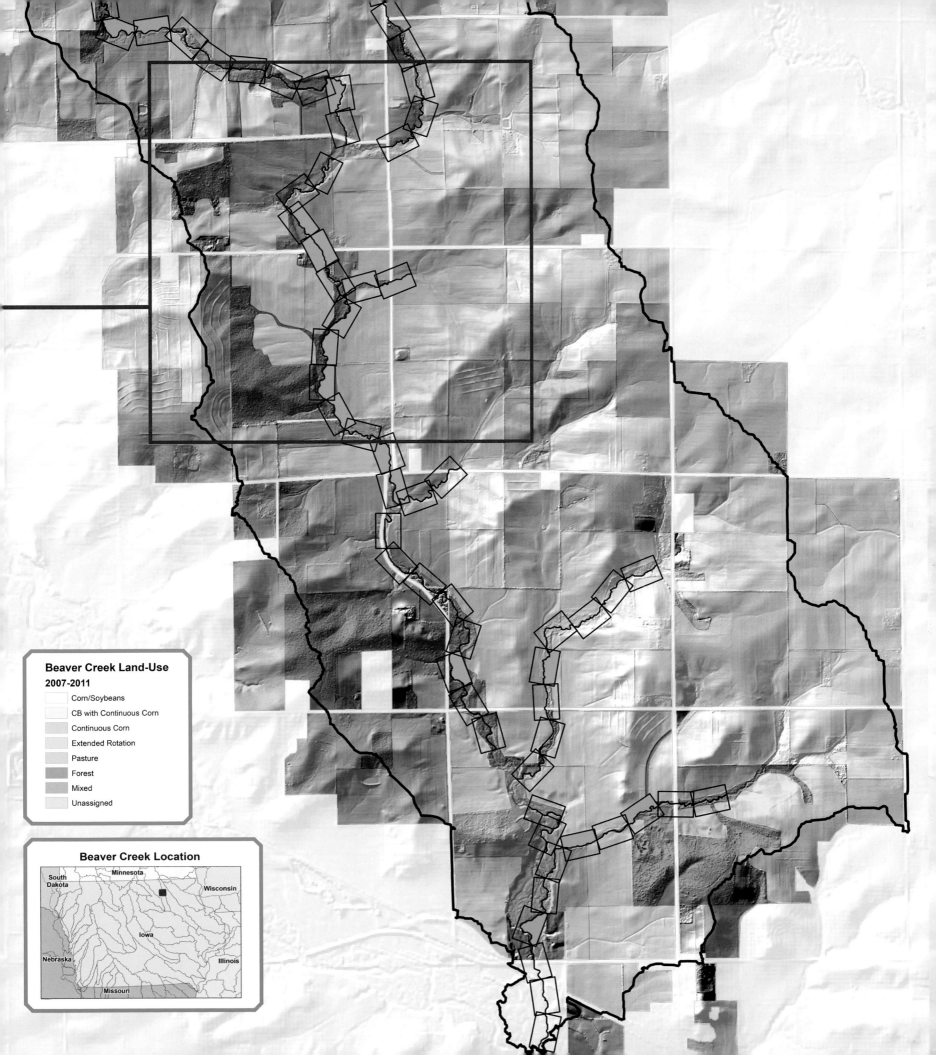

Beaver Creek Land-Use

2007-2011

- Corn/Soybeans
- CB with Continuous Corn
- Continuous Corn
- Extended Rotation
- Pasture
- Forest
- Mixed
- Unassigned

Beaver Creek Location

South Dakota

Minnesota

Wisconsin

Nebraska

Iowa

Illinois

Missouri

Protected Forests in the Amazon

World Wide Fund for Nature
Berlin, Germany

By Mario Barroso and Mariana Soares, WWF-Brazil, and Aurélie Shapiro, WWF-Germany

Contact

Aurélie Shapiro, Aurelie.shapiro@wwf.de

Software

ArcGIS 10.1 for Desktop

Data Sources

Amazon Region Protected Areas Program, World Wide Fund for Nature, Digital Chart of the World, NASA's Moderate Resolution Imaging Spectroradiometer

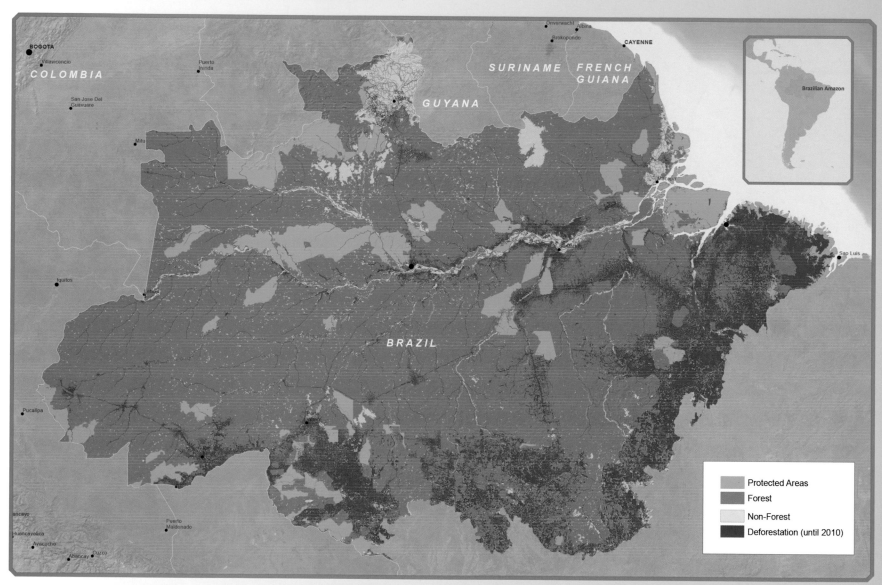

Protected Areas
Forest
Non-Forest
Deforestation (until 2010)

Amazon Region Protected Areas (ARPA) Program is a Brazilian government program led by the Ministry of the Environment and managed by the Brazilian Biodiversity Fund. The program was launched in 2002, designed to last for thirteen years, and implemented in three independent and continuous phases. It is one of the largest tropical forest conservation programs in the world and the biggest one regarding protected areas conservation in Brazil.

ARPA was created with the goal of expanding and strengthening the Brazilian National System of Protected Areas in the Amazon through the protection of 60 million hectares and ensuring financial resources for the management of those areas in the short and long run, while promoting sustainable development in that region.

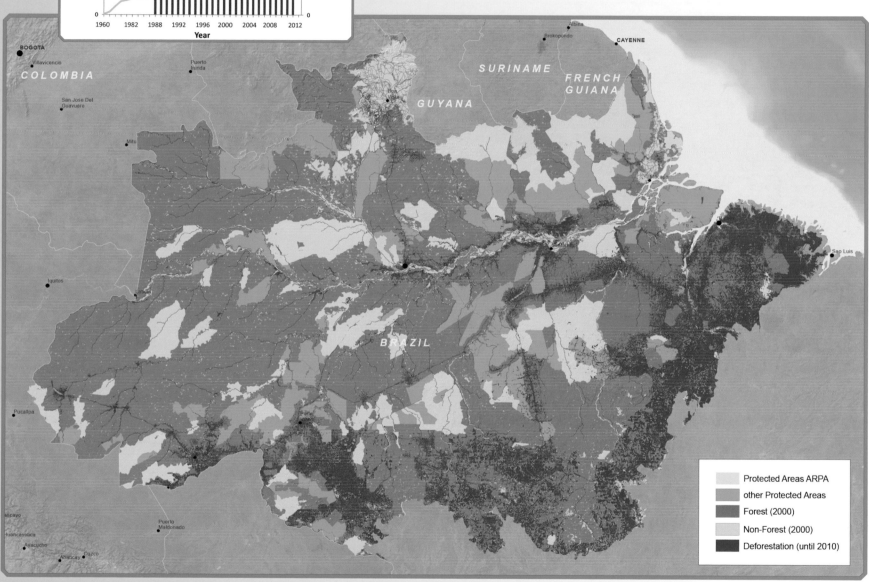

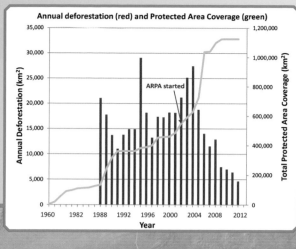

Annual deforestation (red) and Protected Area Coverage (green)

The program, which has earned international recognition, combines conservation biology with best practices in planning and management. Protected areas (PAs) with ARPA support benefit from assets, works, and services, which are needed for the implementation of integration activities involving the communities in the PA surroundings, board creation, management plans, land tenure surveys, patrolling, and other actions necessary for the PA operation.

As shown in these maps, the implementation of ARPA in 2003 was followed by a significant reduction in annual deforestation starting in 2005. Very little deforestation prior to 2010 is observed in ARPA-protected areas demonstrating the efficacy of the program, which is closely aligned with Brazilian government policies and conservation strategies.

Courtesy of Amazon Region Protected Areas Program, World Wide Fund for Nature.

African Elephants: Current and Future Human Impacts

Nordpil
Stockholm, Sweden

By Hugo Ahlenius

Contact
Hugo Ahlenius, hugo.ahlenius@nordpil.com

Software
ArcGIS for Desktop, Adobe Illustrator

Data Source
Natural Earth

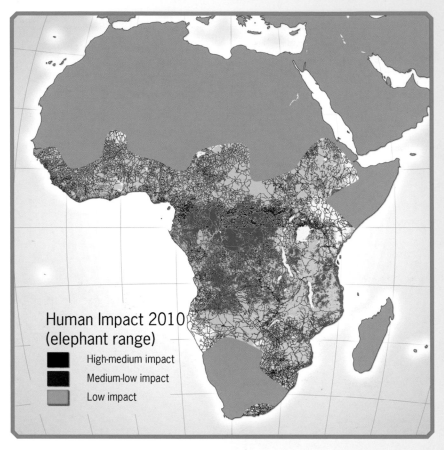

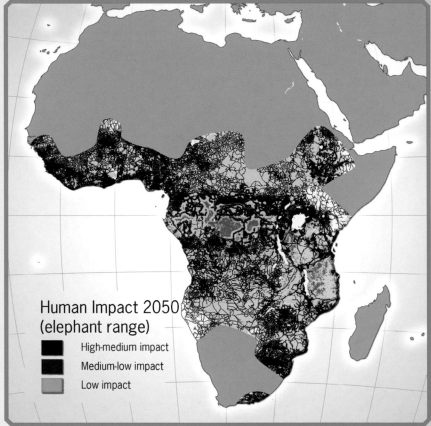

Elephants in the Dust—The African Elephant Crisis is an eighty-page report that provides an overview of the current state of the African elephant alongside recommendations for action to ensure its protection.

For a spatial analysis of the current and future impact of human activities in the current elephant range, the report's authors relied on Nordpil to provide an analysis using the GLOBIO model. The GIS model utilized is based on existing land cover; population; economic activities; and, most importantly infrastructure, such as roads, railroads, and power lines.

Courtesy of Hugo Ahlenius, Nordpil.

Naval Air Station Pensacola Storm Surge

Naval Facilities Engineering Command

Jacksonville, Florida, USA

By Cory Mara

Contact

Cory Mara, cory.mara.ctr@navy.mil

Software

ArcGIS Desktop 9.3.1

Data Sources

Florida Division of Emergency Management,
Florida State Emergency Response Team

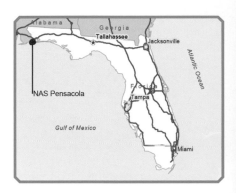

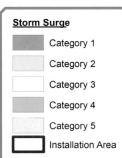

Storm Surge

- Category 1
- Category 2
- Category 3
- Category 4
- Category 5
- Installation Area

The GeoReadiness Center within the Naval Facilities Engineering Command (NAVFAC) Southeast provides geospatial information and services support to multiple Navy installations across the Southeast United States and the Caribbean. This map illustrates hurricane surge potential for category 1 through 5 storms at Naval Air Station Pensacola. The term *storm surge* refers to the rise of water associated with the storm. Potential storm surge factors can include barometric pressure, bathymetry, coastal land formation, elevation, precipitation, tides, wind speed, and wind direction. Category 1 areas (shown in red) are most at risk of flooding during any storm surge event. Although the probability of a category 5 storm (shown in purple) is low, the impacts are far reaching and more severe.

Courtesy of NAVFAC Southeast GeoReadiness Center.

United Nations Peacekeeping Operations Deployment Maps

United Nations

New York, New York, USA

By UN Cartographic Section

Contact

Jeffrey Bliss, blissj@un.org

Software

ArcGIS 10.2 for Desktop, Adobe CS 6, Avenza MAPublisher 9.3

Data Source

Deployment data from United Nations Department of Peacekeeping Operations

These three maps are examples of United Nations peacekeeping operations deployment maps produced by the UN Cartographic Section for inclusion in the UN Secretary General's reports to the UN Security Council. UNTSO, MONUSCO, and MINUSMA are acronyms for United Nations Truce Supervision Organization located in the Middle East, United Nations Organization Stabilization Mission in the Democratic Republic of the Congo, and United Nations Multidimensional Integrated Stabilization Mission in Mali, respectively. These deployment maps depict uniformed personnel (troops, formed police units, and military observers) deployment status in each peacekeeping mission at the indicated period. They are the official maps for the Secretary General's reports. These deployment maps are also used during various briefings. The deployment maps become public once produced and are being updated regularly based on requirements, usually as mission mandates are revised or renewed by the Security Council.

Copyright United Nations 2014.

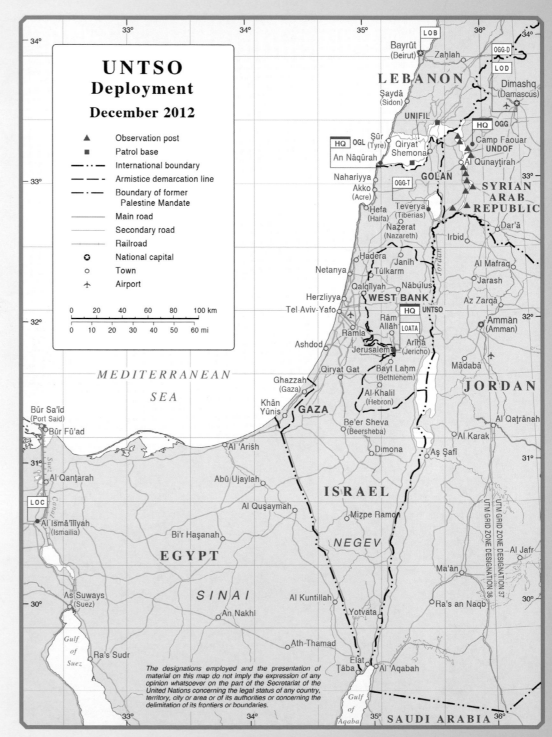

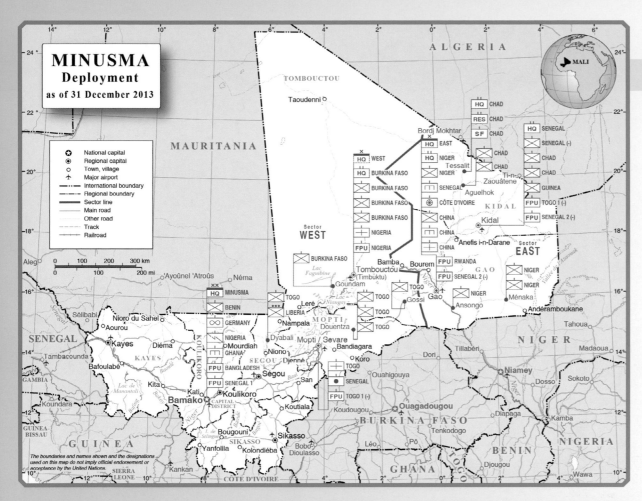

Brownfield Site Analysis

City of Yakima

Yakima, Washington, USA

By Tom Sellsted and Joan Davenport

Contact

Tom Sellsted, Tom.Sellsted@YakimaWA.gov

Software

ArcGIS 10.1 for Desktop

Data Source

City of Yakima

The City of Yakima is evaluating options for remediation and redevelopment of a former landfill, as well as the entire 225-acre Cascade Mill site. The 28-acre landfill area is planned as a future cleanup site under the Environmental Protection Agency (EPA) Brownfield Grant Program. Approximately 408,500 cubic yards of municipal solid waste are present in the former landfill location. The waste was covered by wood waste (average thickness of 3.9 feet) and incorporated into log decks.

The municipal landfill operated from 1965 to 1972, when Interstate 82 was constructed along the Yakima River, which effectively created a river dike and cut off the log ponds from the river. Boise Cascade Mill was fully operational on this site for over 100 years with a lumber mill, a plywood plant, log ponds, kiln buildings, and various associated land uses.

The map data was derived from a 2009 field study by SLR International Corporation that included forty-one soil borings with an average depth of 20 feet; fifty-six test pits; five temporary monitoring wells; and ten soil vapor probes installed to assess combustible gas, which identified methane. Isopach contours of the soil boring data were developed by SLR to identify the limits of municipal solid waste and depth of materials. The layers and thickness of materials are represented for each soil boring location, as well as whether groundwater was found. A cross-section profile of the municipal waste is illustrated.

Courtesy of City of Yakima, Washington, GIS Services.

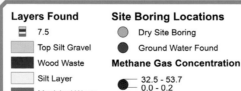

Layers Found

	7.5
	Top Silt Gravel
	Wood Waste
	Silt Layer
	Municipal Waste
75%	Percent of Soil in Municipal Waste by Site Boring Locations

Site Boring Locations

- Dry Site Boring
- Ground Water Found

Methane Gas Concentration

- 32.5 - 53.7
- 0.0 - 0.2
- Waste Area
- Isopach Contours
- Surface Profile Line

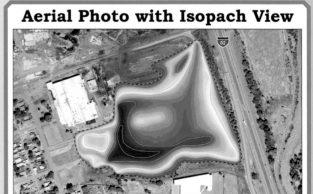

Aerial Photo with Isopach View

Wind Gust Estimates during Superstorm Sandy

New Jersey Department of Environmental Protection

Trenton, New Jersey, USA

By Nick Procopio

Contact

Nick Procopio, nick.procopio@dep.state.nj.us

Software

ArcGIS 10.0 for Desktop

Data Sources

New Jersey Department of Environmental Protection, New Jersey Forest Fire Service

Superstorm Sandy moved across New Jersey October 29–30, 2012. Maximum wind gust speed (knots) was estimated for the period when the storm moved across the region. Data collected at twenty-four weather monitoring stations in New Jersey and twenty-five additional stations neighboring the state was used to estimate the regional wind gusts. Prediction estimates very closely matched the reported value at the forty-nine stations. The average difference was -0.16 percent, while the greatest single deviation was never more than 8.6 percent.

Courtesy of Nick Procopio, New Jersey Department of Environmental Protection 2012.

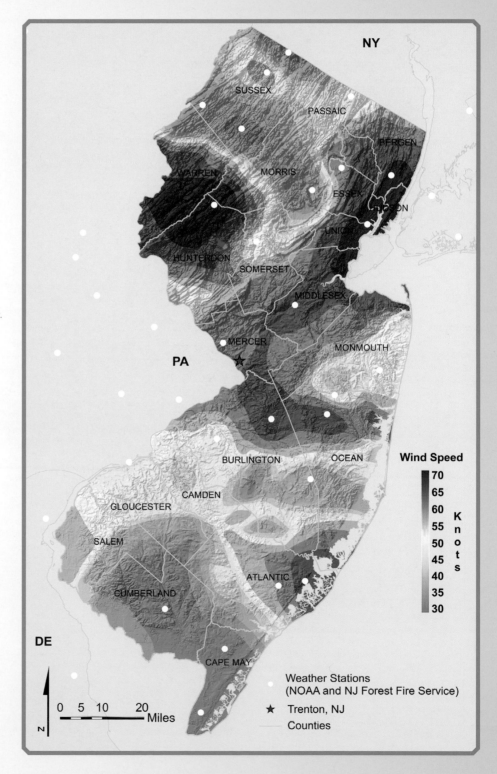

Status of the US Important Bird Areas Program

National Audubon Society

Willow Grove, Pennsylvania, USA

By Tom Auer

Contact

Tom Auer, tauer@audubon.org

Software

ArcGIS 10.1 for Desktop, Adobe Illustrator CS5, Python, RStudio

Data Sources

National Audubon Society, Natural Earth Data, eBird

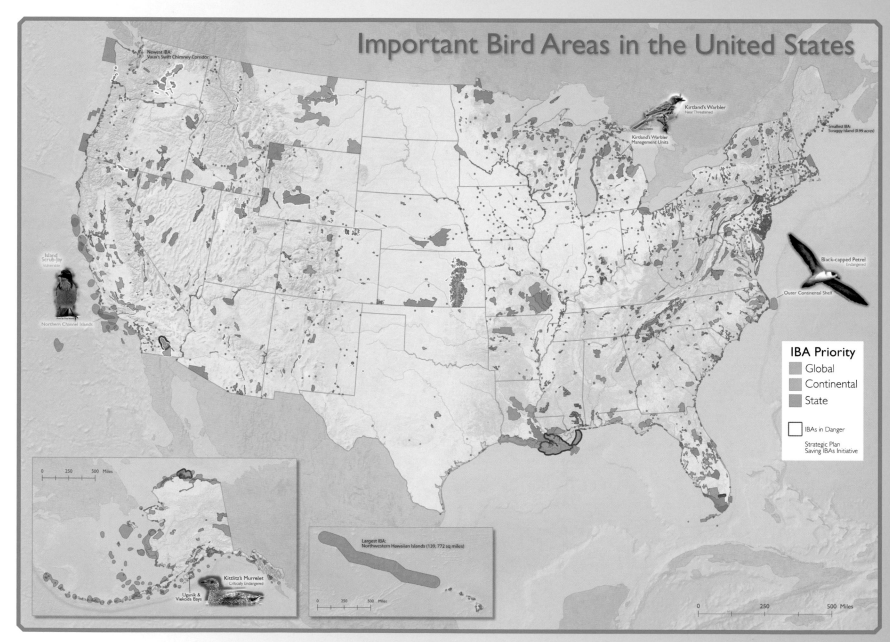

Important Bird Areas in the United States

The Important Bird Areas (IBA) program is an international effort to identify, conserve, and monitor a network of sites that provide essential habitat for bird populations. This map depicts the current state of the IBA program in the United States, showing all public sites coded by their priority, or importance, as determined by the data supporting species that use the site. Global sites often support either globally threatened species or a high percentage of a species' global population. On the map, statistics about the National Audubon Society's network of sites, as well as history and background information about prioritization initiatives, provide context.

Courtesy of National Audubon Society Inc.

Kirtland's Warbler
Near Threatened

Photo by USFWS

Kirtland's Warbler
Management Units

Climate Maps and Data Products for New Zealand

National Institute of Water and Atmospheric Research
Wellington, New Zealand

By James Sturman, Andrew Tait, and John Sansom

Contact

James Sturman, james.sturman@niwa.co.nz

Software

ArcGIS 10.1 for Desktop

Data Sources

Land Information New Zealand (LINZ), National Institute of Water and
Atmospheric Research

The National Institute of Water and Atmospheric Research
(NIWA) National Climate Database holds a vast amount of data
collected at climate stations at specific locations throughout New
Zealand. This data is useful for site-specific climate analysis. For
some applications, such as identifying areas suitable for growing
particular crops, assessing solar and wind power potential, or
comparing very dry years with the long-term average, maps
and spatial datasets are more useful. To better accommodate
these investigations, climate map and data products have been
developed using ANUSPLIN spatial interpolation methodologies,
data extracted from NIWA's National Climate Database, and
ArcGIS 10.1 for Desktop.

New Zealand national and regional annual, seasonal, and
monthly climate maps have been produced for the median, as well
as the twentieth and eightieth percentile statistics, for the 30-year
climate normal period 1981–2010 for various climate variables.
Over 10,000 maps were produced, and a sample of these maps is
shown here.

Courtesy of National Institute of Water and Atmospheric Research, New Zealand.

Ice Shelf Abbot (2002–2012)

Community College of Baltimore County
Catonville, Maryland, USA

By Olesia Ivanenko and Vladimir Belkin

Contact
Scott Jeffrey, sjeffrey@ccbcmd.edu

Software
ArcGIS 10.0 for Desktop, Image Analysis, ERDAS IMAGINE

Data Sources
Esri, National Oceanic and Atmospheric Administration

The Abbot Ice Shelf is a thick slab of ice, attached to West Antarctica's central coast and extending out to the ocean as a seaward extension of the grounded Antarctic Ice Sheet. This ice shelf is one of forty-seven named Antarctic ice shelves. Using ArcGIS for Desktop, Image Analysis, and ERDAS IMAGINE, this project evaluated and compared the position and thickness of the Abbott Ice Shelf in 2002 versus 2012.

An image difference was then conducted to determine the location and amount of change between the two datasets. Then, using wind and current data for the region, along with the results of the image difference, a prediction of the location, extent, and depth of the ice sheet was made for 2022.

Courtesy of Community College of Baltimore County Geospatial Applications Program.

Correlation of High-Pollution Areas and the Possible Effects on California Sea Lion Pups

Cal Poly University, Pomona

Pomona, California, USA

By Everardo Arrizon, Mary Cadena, Cynthia Casarez, and Rose Le

Contact

Lin Wu, lwu@csupomona.edu

Software

ArcGIS 10.0 for Desktop

Data Sources

US Census, Heal the Bay Report, US Geological Survey, National Renewable Energy Laboratory, OpenEI, the Pacific Marine Mammal Center, National Geospatial-Intelligence Agency, National Aeronautics and Space Administration, CGIAR, Southern California Association of Governments

Since January 2013, the Southern California coast had experienced what marine biologists call an Unusual Mortality Event. By April, 1,293 sea lion pups had washed ashore extremely malnourished and dehydrated. It was speculated that pollution coming from heavily populated coastal areas and a decrease in their food supply—anchovy and sardine spawning areas off the coast of California—could be the factors. However, the problem had not been explained definitively.

This project explored the possible contributing factors to this event from spatial and locational perspectives. A weighted overlay model was developed to identify high-population areas and potential pollution sources from different land uses and drainage patterns in the area. Although the modeled high-pollution area sources did not fully correlate with the beach grades from the Heal the Bay Report Card, the maps showed that the spatial correlations of both ocean and land potential pollution sources, the anchovy population distribution, and the dynamics of pollution migration patterns along the Southern California coastal zone offered some explanation of this Unusual Mortality Event.

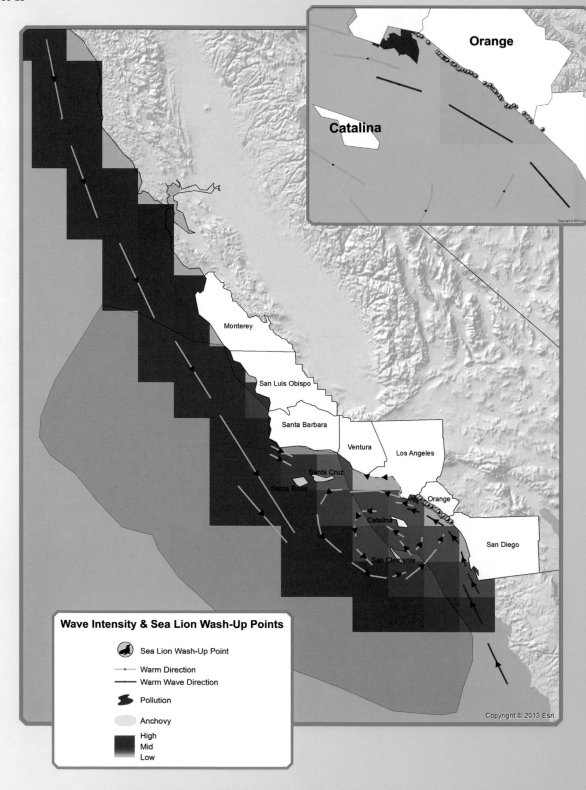

Wave Intensity & Sea Lion Wash-Up Points

- Sea Lion Wash-Up Point
- Warm Direction
- Warm Wave Direction
- Pollution
- Anchovy
- High
- Mid
- Low

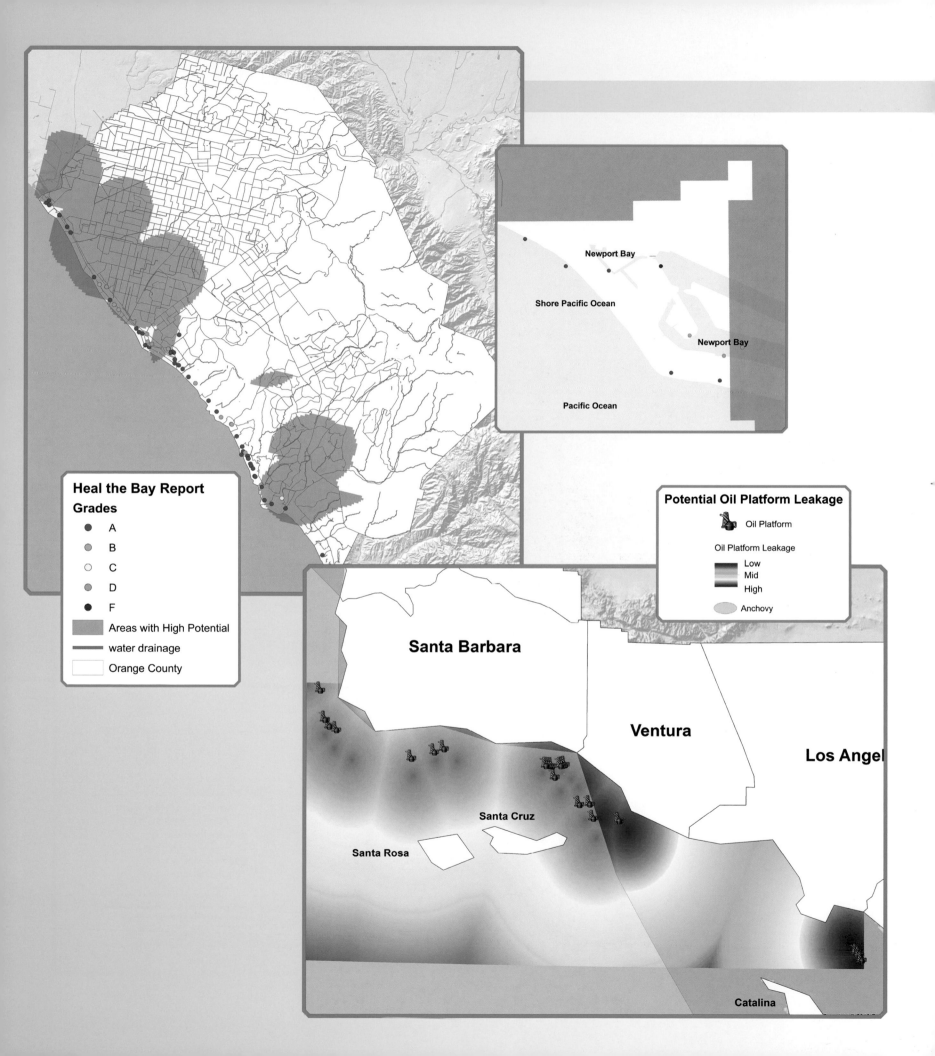

Heal the Bay Report
Grades
- ● A
- ● B
- ○ C
- ● D
- ● F

▬ Areas with High Potential
— water drainage
☐ Orange County

Newport Bay

Shore Pacific Ocean

Newport Bay

Pacific Ocean

Potential Oil Platform Leakage

⚙ Oil Platform

Oil Platform Leakage

Low
Mid
High

⬭ Anchovy

Santa Barbara

Ventura

Los Angel

Santa Cruz

Santa Rosa

Catalina

Projected Usable Lifetime of the Ogallala Aquifer

Texas Tech University

Lubbock, Texas, USA

By Kevin Mulligan, Lucia Barbato, and Santosh Seshadri

Contact

Kevin Mulligan, kevin.mulligan@ttu.edu

Software

ArcGIS 10.1 for Desktop, ArcGIS Spatial Analyst

Data Source

US Geological Survey's Nebraska Water Science Center

The Ogallala Aquifer supports one of the most agriculturally productive regions of the world. Throughout the Great Plains, water drawn from the aquifer is widely used to support irrigated agriculture, livestock production, and rural communities. So how long will the aquifer last? The answer to that question depends on where you are. Beneath the Nebraska Sand Hills, the saturated thickness of the aquifer is more than one thousand feet. In other parts of the region, the aquifer has already been depleted.

To better understand the aquifer, these maps illustrate the spatial variability in saturated thickness, water-level change, and the projected usable lifetime of the aquifer. To develop the maps, saturated thickness and water-level change rasters were obtained from the US Geological Survey's Nebraska Water Science Center.

To estimate the projected usable lifetime of the aquifer, the saturated thickness was divided by the average rate of water-level change. When the saturated thickness is drawn down to about 30 feet, the aquifer can no longer support large-volume irrigation and the aquifer is considered to be effectively depleted.

In many parts of the region, the aquifer is already below 30 feet. Moreover, the results of this analysis suggest that the usable lifetime of the aquifer is on the order of thirty to forty years in parts of Texas, western Kansas, and eastern Colorado. If current rates of aquifer drawdown continue into the future, by 2050 there is likely to be a significant decline in irrigated agriculture on the Great Plains.

Courtesy of Center for Geospatial Technology.

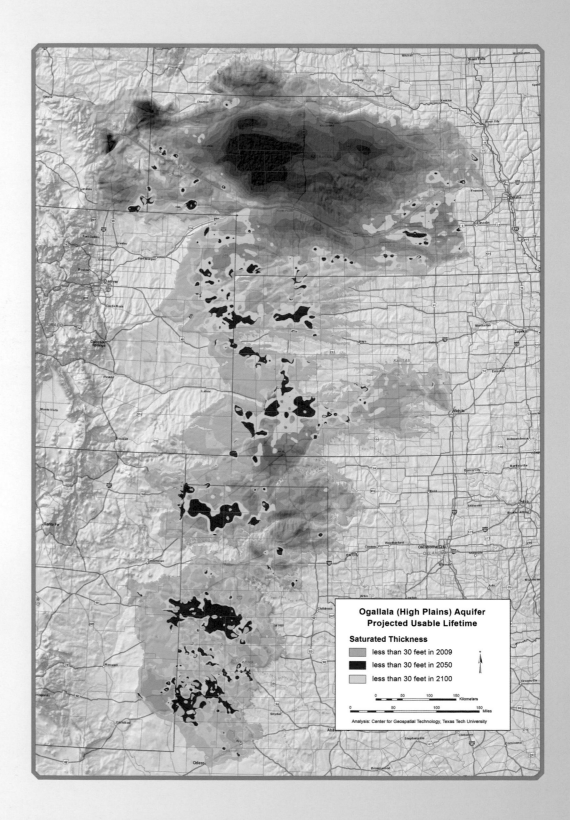

Ogallala (High Plains) Aquifer Projected Usable Lifetime

Saturated Thickness

less than 30 feet in 2009
less than 30 feet in 2050
less than 30 feet in 2100

Analysis: Center for Geospatial Technology, Texas Tech University

Fraudscape: Depicting the United Kingdom's Fraud Landscape

Ordnance Survey

Southampton, Hampshire, United Kingdom

By Christopher Wesson and Paul Naylor

Contact

Christopher Wesson, christopher.wesson@ordnancesurvey.co.uk

Software

ArcGIS 10.0 for Desktop, Adobe Illustrator CS6

Data Source

CIFAS statistics

CIFAS, the United Kingdom's fraud prevention service, in March 2013 released Fraudscape, a report analyzing frauds recorded in 2012. Nearly 250,000 frauds, more than any previous year, were recorded, and identity-related crimes accounted for 65 percent of all fraud confirmed in the United Kingdom that year.

CIFAS conducts its geographic analysis with Ordnance Survey, which also provides CIFAS with a cartographic service ensuring that all maps fit the style of the report, correctly represent the point being made, and match up with the figures produced internally by CIFAS.

The maps shown here display the national picture of fraud for the categories of total frauds, impersonation (identity fraud), and takeover. Takeover is either facility takeover fraud or account takeover fraud. It occurs when a person (the facility hijacker) unlawfully obtains access to information about an account holder or policyholder and fraudulently operates the account or policy for his or her own (or someone else's) benefit. The color scheme used for CIFAS' own maps was already very clear and consistent, so Ordnance Survey chose to adopt the same palette with complementary colors and tie its work in with what had already been produced by the CIFAS graphic designers.

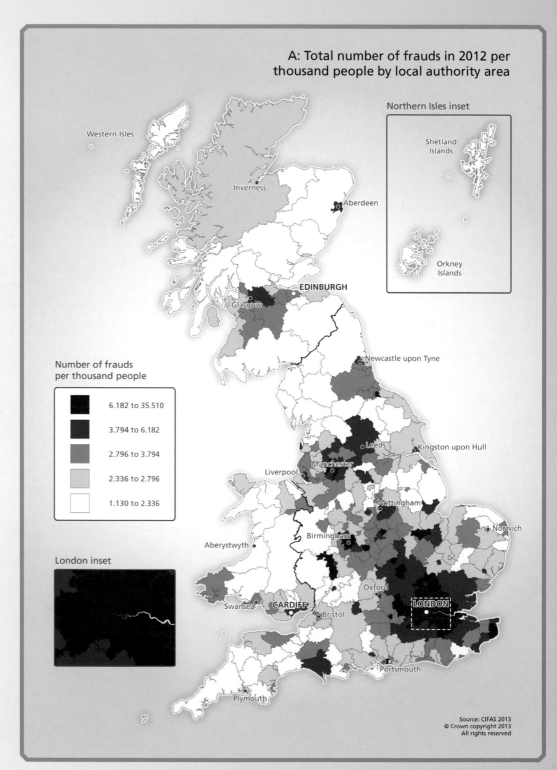

A: Total number of frauds in 2012 per thousand people by local authority area

Northern Isles inset

Number of frauds per thousand people

- 6.182 to 35.510
- 3.794 to 6.182
- 2.796 to 3.794
- 2.336 to 2.796
- 1.130 to 2.336

London inset

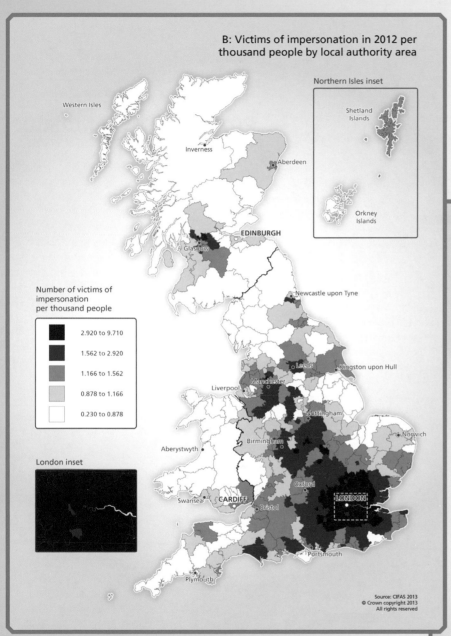

B: Victims of impersonation in 2012 per thousand people by local authority area

Northern Isles inset

Western Isles

Inverness

Aberdeen

Shetland Islands

Orkney Islands

EDINBURGH

Glasgow

Newcastle upon Tyne

Number of victims of impersonation per thousand people

■	2.920 to 9.710
■	1.562 to 2.920
■	1.166 to 1.562
■	0.878 to 1.166
□	0.230 to 0.878

Leeds

Kingston upon Hull

Liverpool

Manchester

Nottingham

Norwich

Birmingham

Aberystwyth

Oxford

London inset

Swansea

CARDIFF

Bristol

LONDON

Portsmouth

Plymouth

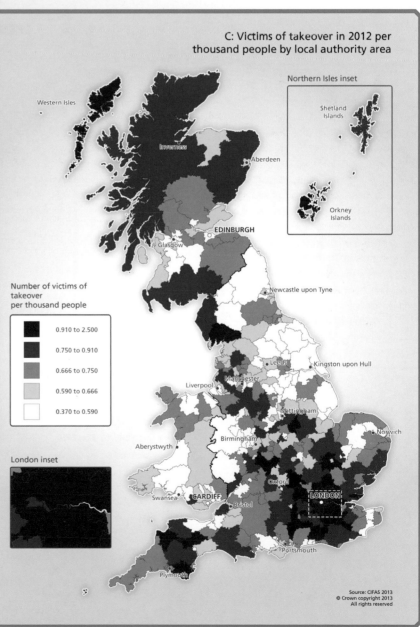

C: Victims of takeover in 2012 per thousand people by local authority area

Northern Isles inset

Western Isles

Inverness

Aberdeen

Shetland Islands

Orkney Islands

EDINBURGH

Glasgow

Newcastle upon Tyne

Number of victims of takeover per thousand people

■	0.910 to 2.500
■	0.750 to 0.910
■	0.666 to 0.750
■	0.590 to 0.666
□	0.370 to 0.590

Leeds

Kingston upon Hull

Liverpool

Manchester

Nottingham

Norwich

Aberystwyth

Birmingham

Oxford

London inset

Swansea

CARDIFF

Bristol

LONDON

Portsmouth

Plymouth

San Francisco Sea-Level Rise and Adaptation Study

URS Corporation

San Francisco, California, USA

By Christian Raumann and Brian Greer

Contact

Brian Greer, brian.greer@urs.com

Software

ArcGIS 10.1 for Desktop

Data Sources

Department of Homeland Security, Port of San Francisco, URS Corporation, City of San Francisco

The Port of San Francisco manages 7.5 miles of shoreline along the city's historic waterfront area. Serving as a landlord for properties located on its wharfs, piers, and landholdings, it's critical for the port to understand the impacts of environmental change and to consider strategies for adaptation. Accordingly, URS Corporation led a multidisciplinary team that integrated scientific understanding, hydrology, civil engineering, and geospatial technologies to estimate the amount, extent, and impact of potential sea-level rise along the waterfront and neighboring inland areas.

These maps illustrate two of the primary measurements of sea-level rise: 100-year flood level (as still water level) and wave run-up (as total water level). Lidar data, imagery, and other survey data were used in a GIS to map and visualize the results generated from the hydrodynamic models MIKE 21 and SWAN for current conditions (2010) and for the years 2050 and 2100. Future scenarios were based on current scientific understanding of sea-level rise. GIS was then used to help the team and stakeholders identify areas of concern and then to conceptually design adaptation measures for each scenario. The results of this study will serve as a guide for the port, waterfront tenants, and developers for project-specific and long-term planning.

Courtesy of URS Corporation, 2013.

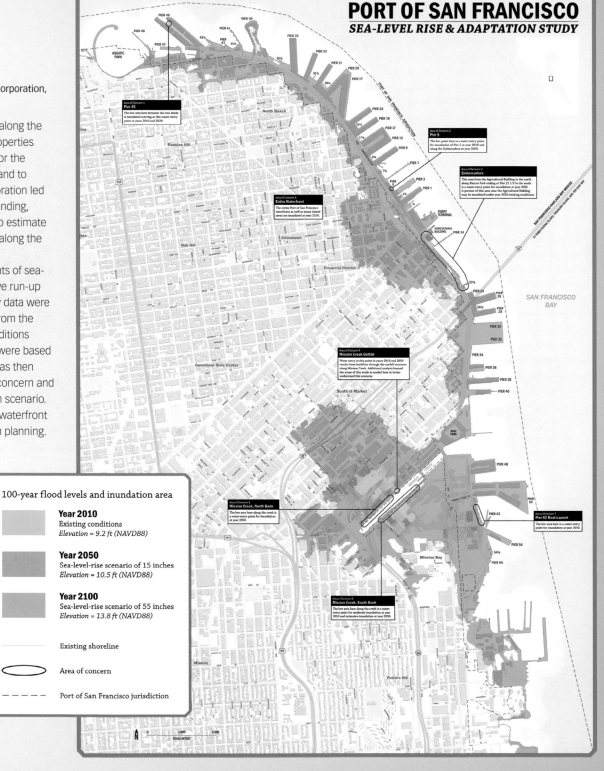

PORT OF SAN FRANCISCO
SEA-LEVEL RISE & ADAPTATION STUDY

100-year flood levels and inundation area

Year 2010
Existing conditions
Elevation = 9.2 ft (NAVD88)

Year 2050
Sea-level-rise scenario of 15 inches
Elevation = 10.5 ft (NAVD88)

Year 2100
Sea-level-rise scenario of 55 inches
Elevation = 13.8 ft (NAVD88)

Existing shoreline

Area of concern

Port of San Francisco jurisdiction

SAN FRANCISCO BAY

PIER 24
PIER 26
26½
PIER 28
PIER 30
PIER 32
PIER 34
PIER 36
PIER 38
PIER 40

Area of Concern 4
Mission Creek Outfall

Water entry at this point in years 2010 and 2050 results from backflow through the outfall structure along Mission Creek. Additional analysis beyond the scope of this study is needed here to better understand this scenario.

South of Market

AT&T PARK

PIER 48

MISSION CREEK

PIER 50

Area of Concern 5
Mission Creek, North Bank

The low area here along the creek is a water-entry point for inundation at year 2050.

PIER 52

Area of Concern 7
Pier 52 Boat Launch

The low area here is a water-entry point for inundation at year 2050.

PIER 54

54½

Mission Bay

CAMPUS WAY

PIER 64

Area of Concern 6
Mission Creek, South Bank

The low area here along the creek is a water-entry point for moderate inundation at year 2010 and extensive inundation at year 2050.

Marsh Creek Dam Inundation Zone

Chester County

West Chester, Pennsylvania, USA

By David Sekkes

Contact

David Sekkes, DSekkes@chesco.org

Software

ArcGIS 10.0 for Desktop, Adobe CS5

Data Sources

Chester County Department of Emergency Services, Chester County GIS

In 2012, Chester County emergency planners reviewed the Marsh Creek Dam Emergency Action Plan (EAP) as part of the five-year review cycle. For the first time, they were able to use a GIS-based inundation zone for analysis. During this review, the team identified the estimated at-risk population found in the plan was surprisingly low. Using the newly acquired GIS inundation zone layer, emergency planners analyzed the at-risk population of the inundation zone based on parcel-level data. The results indicated that the population estimate used in the EAP was off by over 7,000 people.

Then, in early 2013, Chester County used a building footprint data layer to rerun the analysis of the at-risk population of the inundation zone. The new results found that many parcels previously identified as at risk either had no structures or the structures themselves were not truly at risk. The accuracy of the building footprints lowered the number of parcels at risk of flooding from 3,214 (using the parcels) down to 2,138 (using building footprints). The estimated impacted population declined from 8,500 to 6,900. The final population estimate is more than four times greater than what was included in the old plan. Using GIS and building footprints, the county now has a high degree of confidence in the accuracy of the number of people and properties at risk in the current EAP.

Courtesy of Chester County Department of Emergency Services.

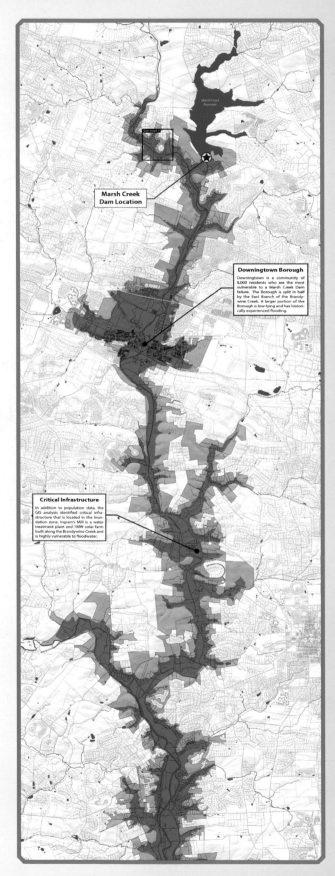

Marsh Creek Dam Location

Downingtown Borough
Downingtown is a community of 8,000 residents who are the most vulnerable to a Marsh Creek Dam failure. The Borough is split in half by the East Branch of the Brandywine Creek. A larger portion of the Borough is low-lying and has historically experienced flooding.

Critical Infrastructure
In addition to population data, the GIS analysis identified critical infrastructure that is located in the inundation zone. Ingram's Mill is a water treatment plant and 1MW solar farm built along the Brandywine Creek and is highly vulnerable to floodwater.

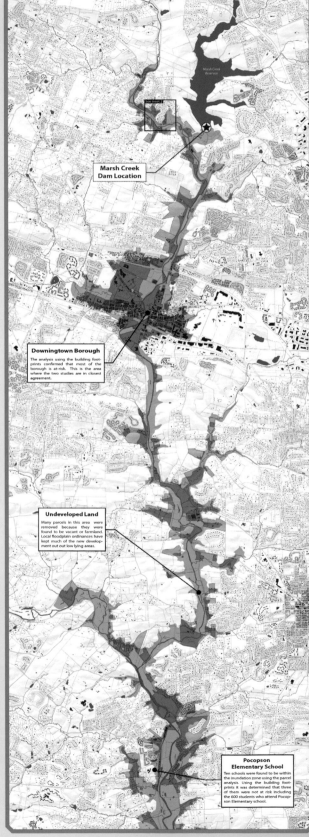

Marsh Creek Dam Location

Downingtown Borough
The analysis using the building footprints confirmed that most of the borough is at-risk. This is the area where the two studies are in closest agreement.

Undeveloped Land
Many parcels in this area were removed because they were found to be vacant or farmland. Local floodplain ordinances have kept much of the new development out of low lying areas.

Pocopson Elementary School
Ten schools were found to be within the inundation zone using the parcel analysis. Using the building footprints it was determined that three of them were not at risk including the 600 students who attend Pocopson Elementary school.

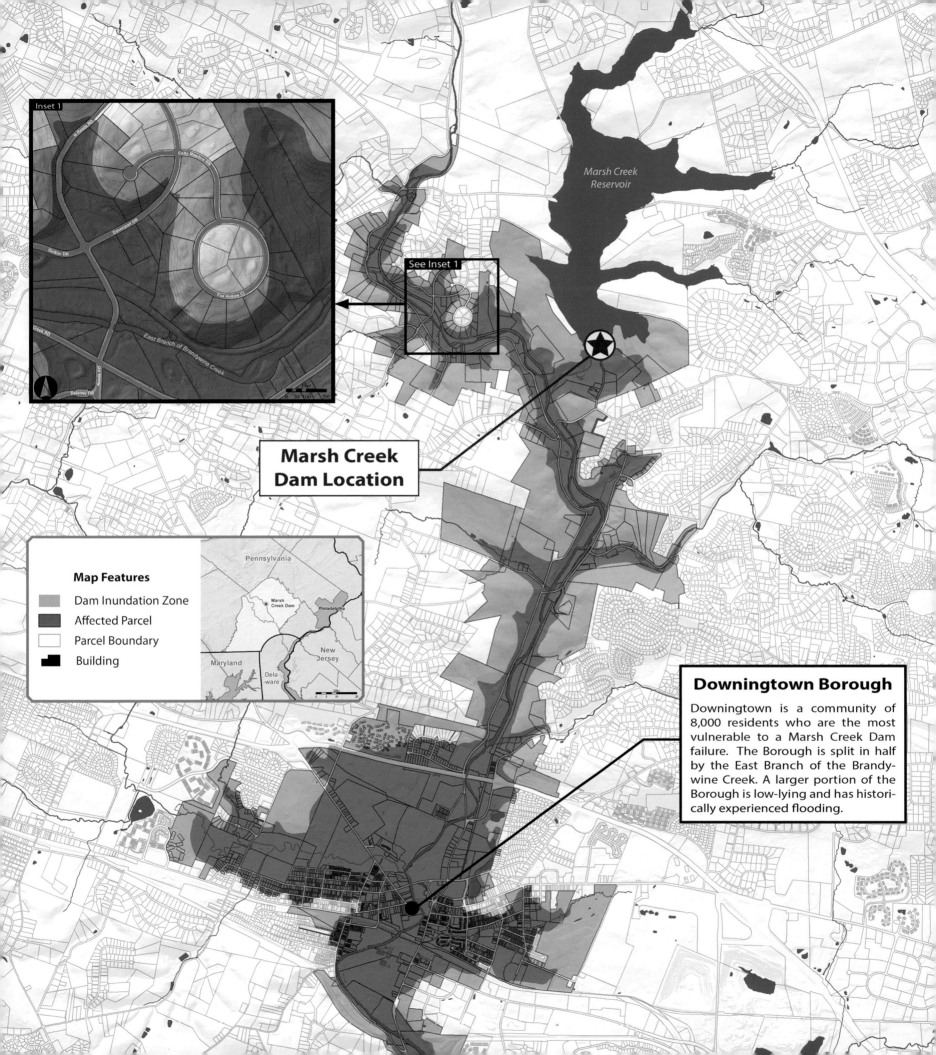

Inset 1

N Reeds Rd
Colts Meadow Rd
Davenport DR
Gomer DR
Creek RD
Fox Hollow Ct
Reeds RD
East Branch of Brandywine Creek
Delaney DR

0 50 100 200 Feet

Marsh Creek
Reservoir

See Inset 1

★ Marsh Creek
Dam Location

Map Features

Dam Inundation Zone
Affected Parcel
Parcel Boundary
Building

Pennsylvania
Marsh Creek Dam ★
Philadelphia
Maryland
New Jersey
Dela-ware

Downingtown Borough

Downingtown is a community of 8,000 residents who are the most vulnerable to a Marsh Creek Dam failure. The Borough is split in half by the East Branch of the Brandy-wine Creek. A larger portion of the Borough is low-lying and has histori-cally experienced flooding.

Inundation Vulnerability Map for Management of Natural Disasters

Daewon Institute

*Hayang-eup, Kyungsan-si,
Gyeongsang, Republic of South Korea*

By Wanyoung Song and Jongbae Kim

Contact

Wanyoung Song, it4korea@naver.com

Software

ArcGIS 10.0 for Desktop, Adobe Photoshop 8.0

Data Source

Vector map of Korea

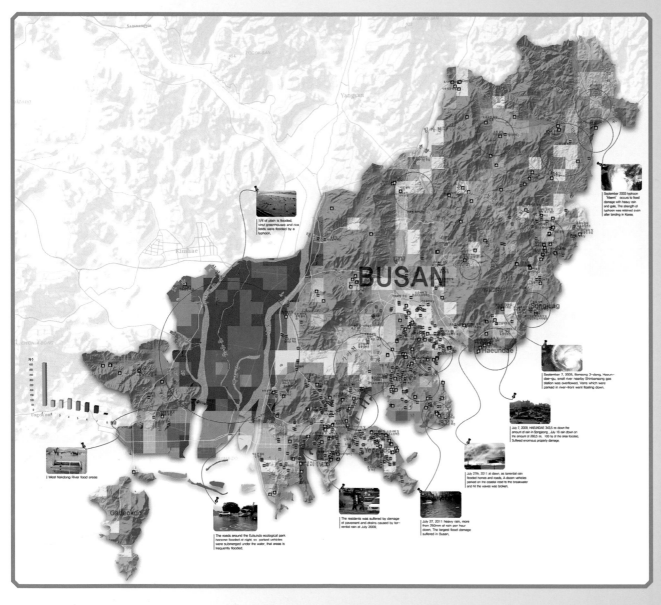

The central government of South Korea's safety management tasks have been simplified with the development of web-based map services. These analysis maps assess risk of flooding caused by natural disaster in the city of Busan, the country's second-largest metropolis after Seoul. Three types of analysis factors—topographic, social factors, and damage mitigation facilities for flood disaster—are selected and analyzed. Disaster prevention information is provided by web-based services by assessing vulnerable flooding areas, providing basic information, and supporting disaster prevention policies. A map magazine service helps users understand the disaster prevention information and enables the service providers to generate and distribute information easily. It is not just a simple map viewer but a public-oriented publishing service of disaster prevention information for maximum utilization.

Courtesy of Wanyoung Song.

Map Indicating Flooding in the Savanna of Bogotá and Its Surroundings

SiGaia

Bogotá, Colombia

By Fernando Salazar-Holguín

Contact

Fernando Salazar-Holguín, fersalazar@sigaia.com

Software

ArcGIS 10.1 for Desktop, ILWIS 3.3 Academic

Data Sources

Shuttle Radar Topography Mission; HydroSHEDS; Instituto de Hidrología, Meteorología y Estudios Ambientales—IDEAM; Instituto Geográfico

The Savanna of Bogotá, a plateau in the central region of Colombia, is an important agricultural region as well as the most industrialized and densely populated area of the country. The savanna contains especially rich ecosystems in floodplains and wetlands, habitat for particular native flora and fauna, as well as critical stopover and wintering grounds for migratory birds. Together with forests, they play a crucial role in the hydrological cycle, both attenuating floods and maintaining flow during dry periods.

While a great part of these natural landscapes has been artificially drained for development, there are also new ones artificially created every day by dams, dikes, and mining excavations. This map shows where flooding will eventually occur due to the variable flow of water, the specific geology, soil and vegetation types, infrastructure, and so on, delineating contours with corresponding depths and volumes. El Niño and La Niña Southern Oscillations (ENSO) and other severe weather phenomena, tsunamis, landslides, and accidents also determine flooding behavior that increasingly causes loss of life and damages.

Courtesy of SiGaia.

			Ptos. Topo.	**Pols. Topo.**	**Curvas 1000 m**		
★ Cap. Deptal.	— Via_Ferrea 100K	⋯ Drenaje_Sencillo	✦ Picos	▣ Picos	— 6000		
★ Cab. Mun. y Corr. Dept.	— Via Tipo 1	Drenaje Doble	× Boquerones	Cuchillas	— 5000		
● Corregimientos	→ Via Tipo 2	Isla	▼ Hondonadas	Boquerones	— 4000		
✇ Arqueología	— Via Tipo 3	Laguna		Encañadas	— 3000		
◿ Parque Nacional Natural	⌐ Via Tipo 4	Pantano		Depresiones	— 2000		
✕ Minas	Via Tipo 5				— 1000		
◿ Áreas urbanas	— Via Tipo 6						
◺ Canteras y minas	— Camino, Sendero						
— Recorridos GPS SiGaia	Peatonal Urbana						

Matices hipsométricos (msnm SRTM)

☐ < 500	☐ 1,201 - 1,300	☐ 2,001 - 2,100	☐ 2,801 - 2,900
☐ 501 - 600	☐ 1,301 - 1,400	☐ 2,101 - 2,200	☐ 2,901 - 3,000
☐ 601 - 700	☐ 1,401 - 1,500	☐ 2,201 - 2,300	☐ 3,001 - 3,100
☐ 701 - 800	☐ 1,501 - 1,600	☐ 2,301 - 2,400	☐ 3,101 - 3,200
☐ 801 - 900	☐ 1,601 - 1,700	☐ 2,401 - 2,500	☐ 3,201 - 3,300
☐ 901 - 1,000	☐ 1,701 - 1,800	☐ 2,501 - 2,600	☐ 3,301 - 3,400
☐ 1,001 - 1,100	☐ 1,801 - 1,900	☐ 2,601 - 2,700	☐ 3,401 - 3,500
☐ 1,101 - 1,200	☐ 1,901 - 2,000	☐ 2,701 - 2,800	☐ > 3,500

2012 Atlantic Hurricane Season

QBE

Irvine, California, USA

By Matthew Eimers

Contact

Matthew Eimers, matthew.eimers@us.qbe.com

Software

ArcGIS 10.1 for Desktop

Data Sources

QBE, National Hurricane Center, National Oceanic and Atmospheric Administration

QBE Insurance Group Limited is one of the top twenty insurers and reinsurers worldwide. Its North American presence focuses on property and casualty, lender-placed, specialty, and crop insurance. The 2012 hurricane season, including the devastating Hurricane Isaac and Superstorm Sandy, significantly affected customers. QBE's CAT Modeling Department uses GIS to map the potential path or footprint of each storm around the clock. QBE provides financial institution clients with live interactive maps of properties located in their portfolios, monitors potential threats in those areas, and estimates the number of claims and destructive costs before landfall. This map is updated on a daily basis to analyze the potential impact before, during, and after landfall.

Courtesy of QBE.

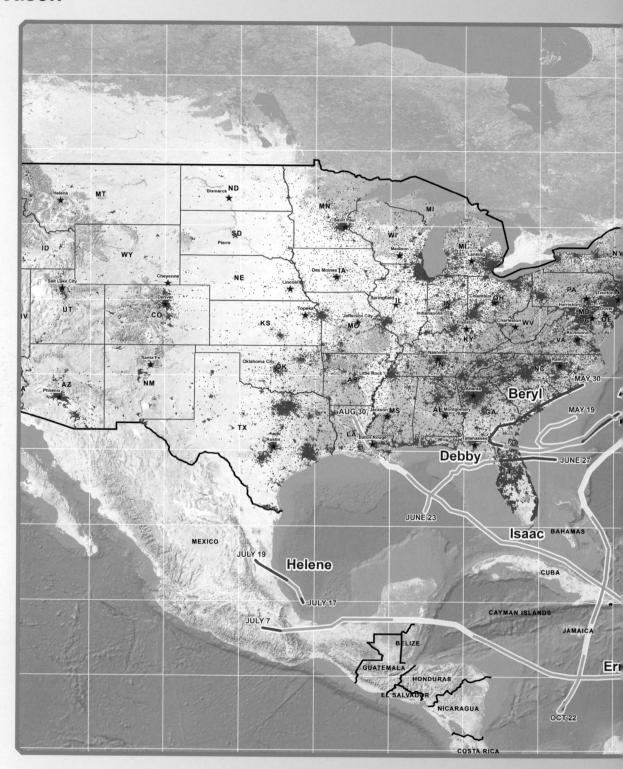

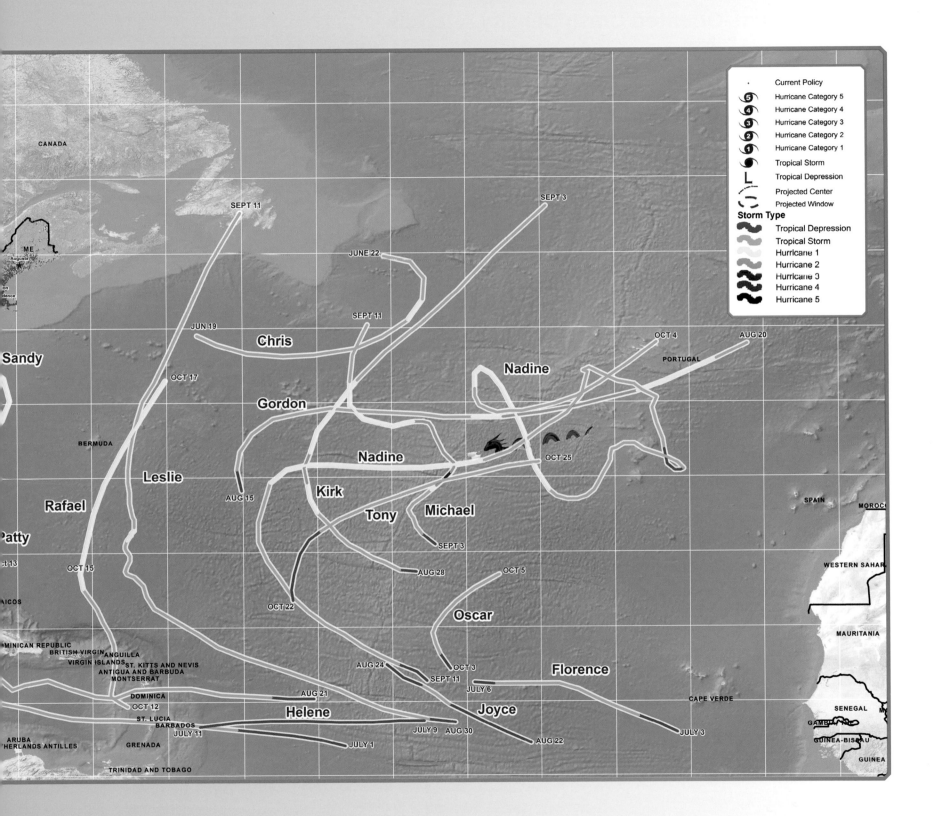

43

Road Segment–Oriented Crash Density Analysis

Virginia Tech
Blacksburg, Virginia, USA

By Joseph Newman

Contact
Joseph Newman, jmano7@vt.edu

Software
ArcGIS 10.1 for Desktop

Data Sources
Virginia Department of Motor Vehicles, Virginia Highway Safety Office, Virginia Department of Transportation

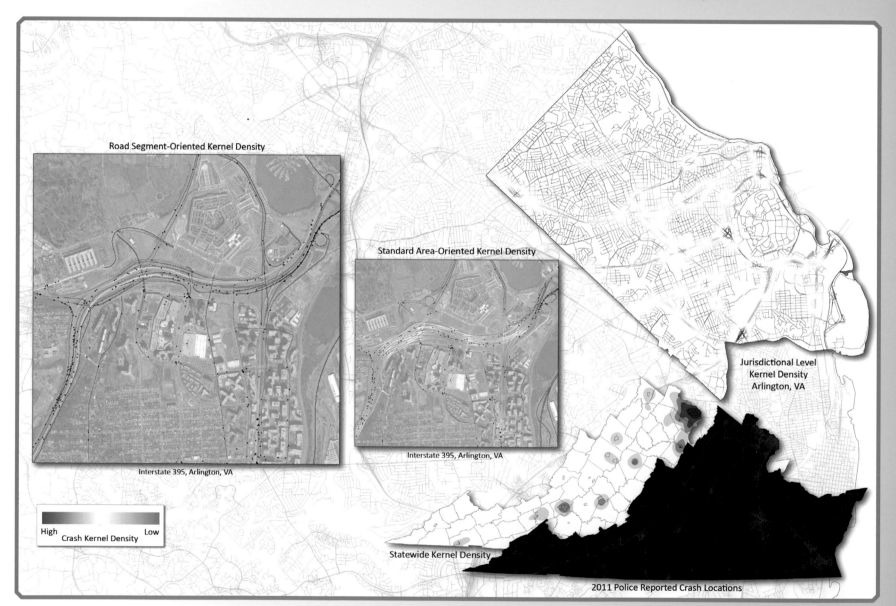

Road Segment–Oriented Kernel Density

Interstate 395, Arlington, VA

Standard Area-Oriented Kernel Density

Interstate 395, Arlington, VA

Jurisdictional Level Kernel Density Arlington, VA

High — Low
Crash Kernel Density

Statewide Kernel Density

2011 Police Reported Crash Locations

This map shows how highway safety officials more accurately identify hazardous areas along the Commonwealth of Virginia's roadways, allowing them to better direct mitigation strategies. It leverages a comprehensive database of located police reported vehicle crashes within the commonwealth, which is a product of the collaboration between the Virginia Department of Motor Vehicles and the Center for Geospatial Information Technology at Virginia Tech. This database has the potential to improve research, analysis, and planning within the field of transportation and highway safety.

Courtesy of Joseph Newman, Center for Geospatial Information Technology, Virginia Tech.

Spatial Analysis of 3-1-1 Calls in Houston

City of Houston
Houston, Texas, USA

By Sona Ann Sunny

Contact
Sona Ann Sunny, -sona.sunny@houstontx.gov

Software
ArcGIS 10.1 for Desktop

Data Sources
City of Houston Enterprise Geodatabase, US Census Bureau

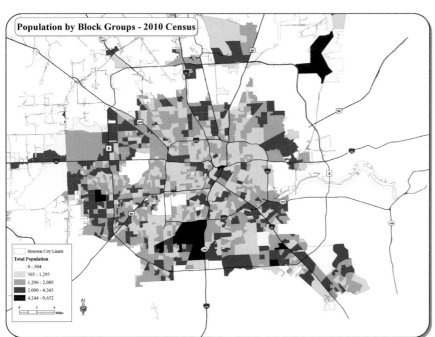

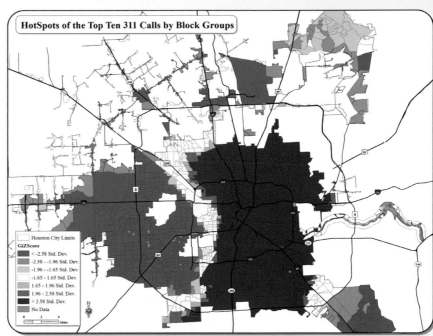

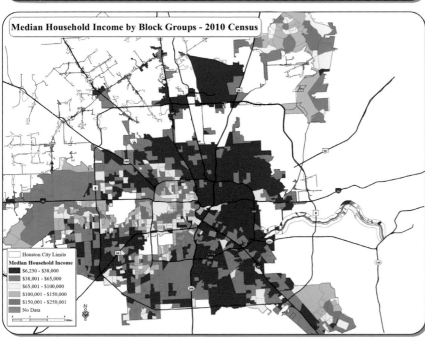

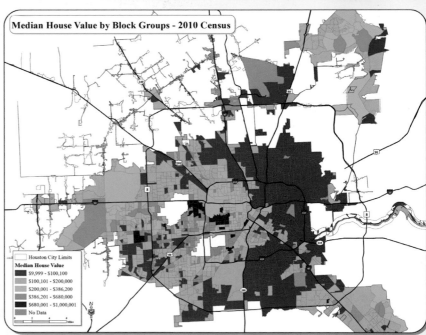

The City of Houston's 3-1-1 Houston Service Helpline, launched in August 2011, enables citizens to report nonemergency concerns, from traffic fines and sewer issues to pothole problems and neighborhood complaints. From November 2011 through December 2012, the city received 216,851 calls with 230 different types of requests.

This map shows the spatial patterns of those 3-1-1 calls and their relationship to the demography of the area. The calls were geocoded, and the top ten requests were extracted for use in a hot spot analysis to explore the spatial relationship of the calls on a citywide level. The z-scores that indicate the statistical significance of the

clustering of calls were mapped to show the hot and cold spots. The calls were then aggregated by 2010 Census block groups, and the demographics were mapped to help understand some of the causes of the higher concentration of calls in the hot spot areas.

Courtesy of City of Houston.

A Small City Approach to Mapping

City of West Linn
West Linn, Oregon, USA

By Kathy Aha, GISP

Contact

Kathy Aha, kaha@westlinnoregon.gov

Software

ArcGIS 10.0 for Desktop

Data Sources

West Linn GIS and Economic Development Department, Metro Regional Land Information System (RLIS), Clackamas County GIS

The City of West Linn, located 10 miles south of Portland, Oregon, with a population of 25,425, has built a robust GIS over the last fifteen years. More than 200 specialized map products and applications are developed each year from a one-person department. The business license map shown here is an example of the city's wide variety of mapping topics that include public planning processes and visioning, economic development, public works, engineering, street maintenance, city administration, and parks.

The business license mapping project utilizes West Linn's business license renewal database and geocoded point locations of all businesses within the city limits. Six business license categories were identified and unique colors are represented on the map to help the Economic Development Committee during meetings and presentations.

Courtesy of City of West Linn.

Legend
2012 Business Licenses*
Total: 668
- Consulting - 217
- Health - 91
- Property - 46
- Retail - 117
- Services - 109
- Other -88

Yellow Highlights:
- Home Occupation Licenses
  349 (52%)

Business Licenses Outside
West Linn City Limits
Included in Totals:
- 10 locations

- Map Numbering Grid
- Metro's UGB, West Linn
- Commercial Centers

GENERAL CATEGORY	CLASSIFICATION	COUNT
Consulting	Attorney/CPA/Accounting Services	29
	Consultant	142
	Contractor	46
Health	Massage Therapy	14
	Medical/Dental	62
	Nursing Home	15
Property	Apartment	24
	Property Manager	6
	Realtor/Commercial Lessor	16
Retail	Amusement	1
	Arts and Crafts	15
	Retail Goods	49
	Retail Grocery	6
	Retail Service	46
Services	Auto Sales/Service	7
	Barber/Beauty Salon	23
	Day Care	8
	Financial Institution	15
	Gas/Service Station	4
	Hotel/Motel	1
	Janitorial/House Cleaning	3
	Landscaping	9
	Restaurant/Lounge	39
Other	Industrial	1
	Manufacturing	7
	Miscellaneous	66
	Nonprofit Org.	3
	Wholesale	11

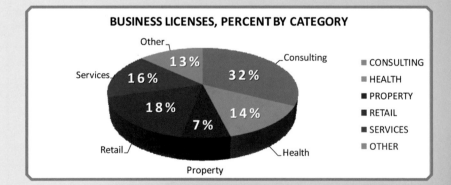

BUSINESS LICENSES, PERCENT BY CATEGORY

Other 13%
Consulting 32%
Services 16%
Retail 18%
Property 7%
Health 14%

- CONSULTING
- HEALTH
- PROPERTY
- RETAIL
- SERVICES
- OTHER

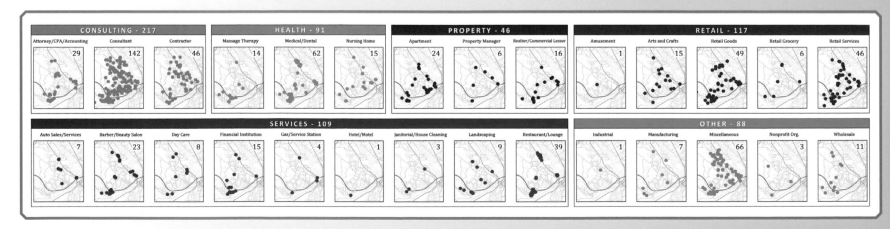

CONSULTING - 217 | Attorney/CPA/Accounting 29 | Consultant 142 | Contractor 46

HEALTH - 91 | Massage Therapy 14 | Medical/Dental 62 | Nursing Home 15

PROPERTY - 46 | Apartment 24 | Property Manager 6 | Realtor/Commercial Lessor 16

RETAIL - 117 | Amusement 1 | Arts and Crafts 15 | Retail Goods 49 | Retail Grocery 6 | Retail Services 46

SERVICES - 109 | Auto Sales/Services 7 | Barber/Beauty Salon 23 | Day Care 8 | Financial Institution 15 | Gas/Service Station 4 | Hotel/Motel 1 | Janitorial/House Cleaning 3 | Landscaping 9 | Restaurant/Lounge 39

OTHER - 88 | Industrial 1 | Manufacturing 7 | Miscellaneous 66 | Nonprofit Org. 3 | Wholesale 11

El Segundo Fire Department Risks and Resources

City of El Segundo

El Segundo, California, USA

By Michael McDaniel

Contact

Michael McDaniel, mmcdaniel@elsegundo.org

Software

ArcGIS 10.1 for Desktop

Data Source

City of El Segundo

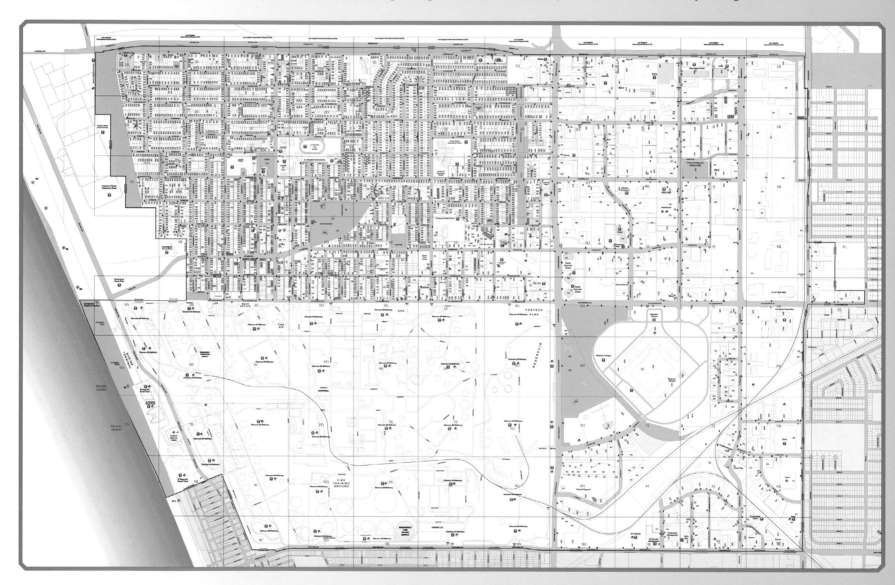

The El Segundo Fire Department has mounted maps in the equipment bays at its stations showing the risks and resources available in the city and surrounding mutual aid areas. Risks include lot configurations (especially extra-deep "flag lots") requiring special resources, hazardous chemical locations, and identified target hazards. Resources include fire hydrants; fire department connections; and draft water sources such as reservoirs, lakes, and even swimming pools. These assets are shown on a basemap showing streets, addresses, major buildings, schools, and so on, which also serves as an index to large-scale map books carried on all fire vehicles.

Courtesy of City of El Segundo.

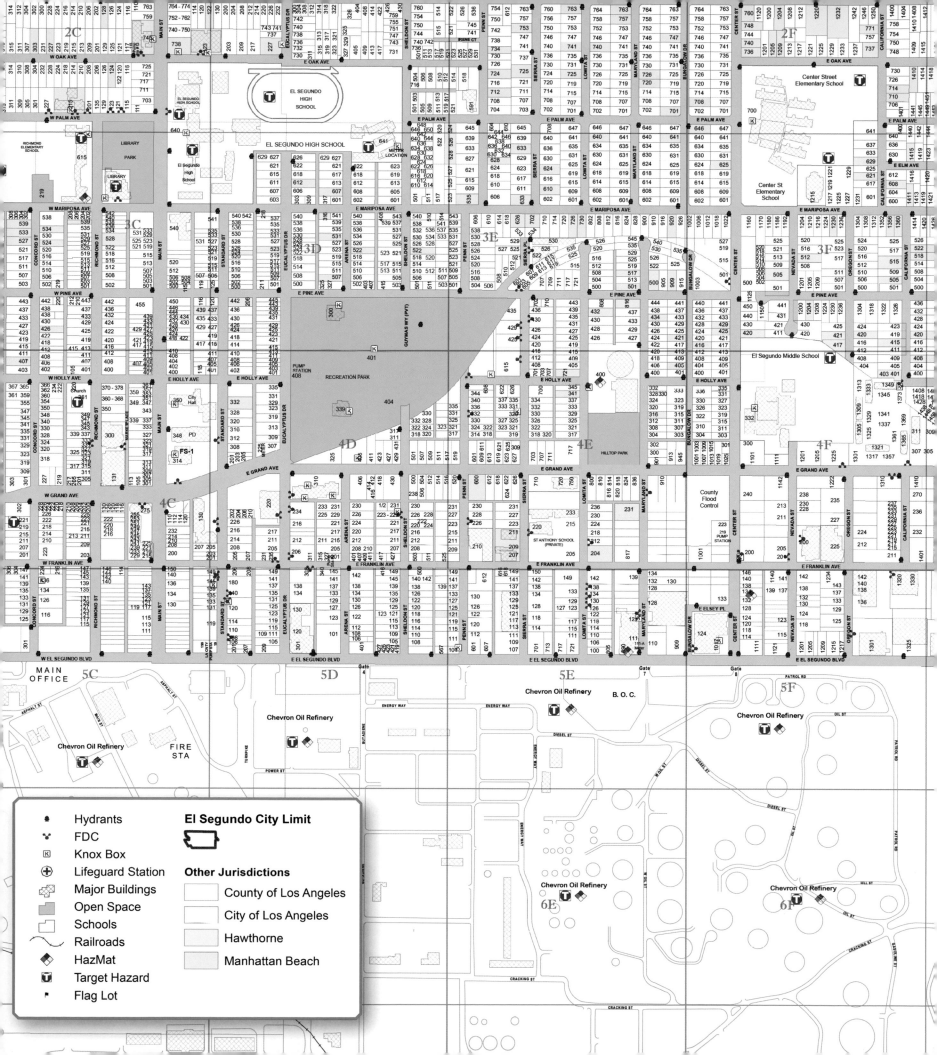

Hydrants
FDC
Knox Box
Lifeguard Station
Major Buildings
Open Space
Schools
Railroads
HazMat
Target Hazard
Flag Lot

El Segundo City Limit

Other Jurisdictions
County of Los Angeles
City of Los Angeles
Hawthorne
Manhattan Beach

Favorability Mapping for New Infrastructure Development— Utilizing Existing Linear Footprint

Government of Alberta

Edmonton, Alberta, Canada

By Tom Churchill, Stéphane La Rochelle, and Elaine Umeris

Contact

Tom Churchill, Tom.Churchill@gov.ab.ca

Software

ArcGIS Desktop 9.3

Data Sources

Energy Resources Conservation Board, IHS Energy, Alberta Electric System Operator

Using spatial analysis techniques, the Alberta Department of Energy examined existing energy infrastructure, specifically linear footprint, to identify areas of high-density infrastructure. Data considered in this analysis included historical seismic exploration cutlines, electrical transmission lines, pipelines, and roads. All features were used to generate a density grid. The high-density areas, or "hot spots," showed favorable locations for new energy infrastructure within or near established areas. Using these areas would help manage the placement of surface disturbance, which, in turn, would lessen environmental impact. These results provided information for the department with respect to developing policy options and recommendations and implementing policy direction in support of the Government of Alberta's integrated resource management system.

Courtesy of the Government of Alberta.

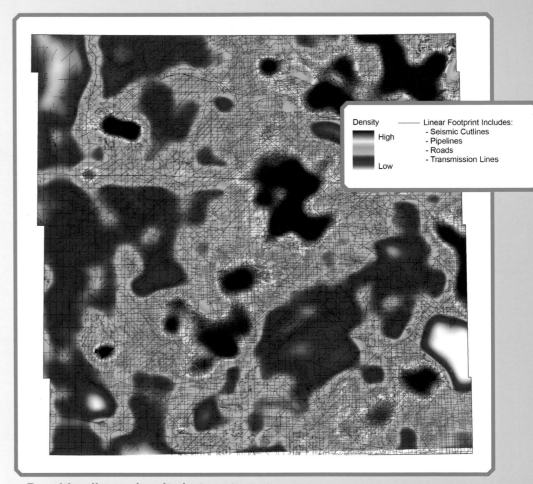

Resulting linear density hot spots

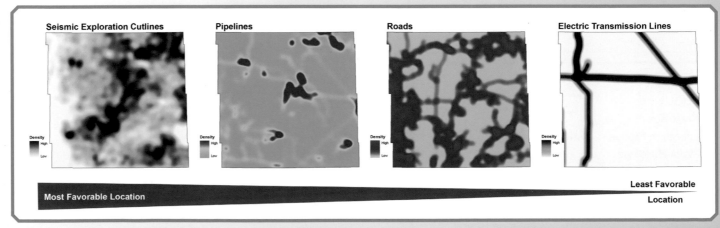

Current linear footprint weighted by favorability

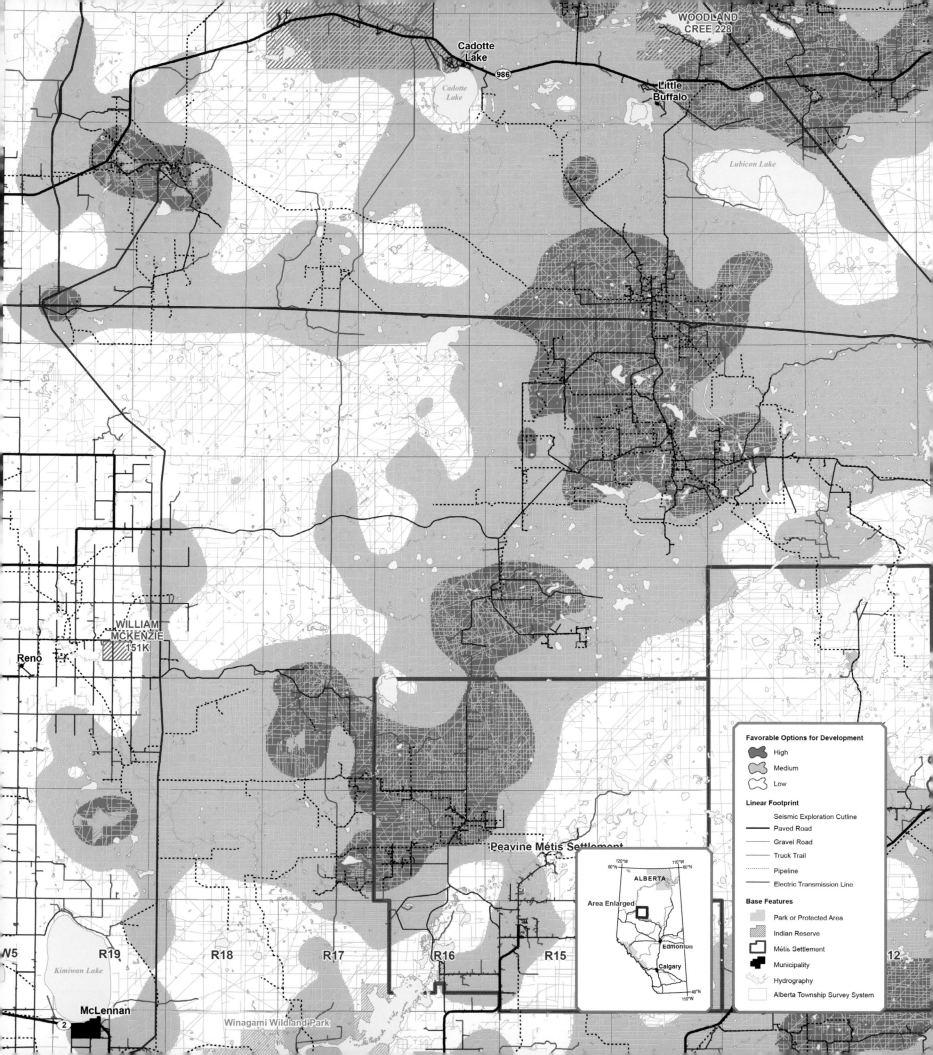

WOODLAND
CREE 228

Cadotte
Lake

Cadotte Lake

986

Little
Buffalo

Lubicon Lake

WILLIAM
MCKENZIE
151K

Reno

Peavine Métis Settlement

McLennan

2

Kimiwan Lake

W5   R19   R18   R17   R16   R15   12

Winagami Wildland Park

Favorable Options for Development

High

Medium

Low

Linear Footprint

Seismic Exploration Cutline

Paved Road

Gravel Road

Truck Trail

Pipeline

Electric Transmission Line

Base Features

Park or Protected Area

Indian Reserve

Métis Settlement

Municipality

Hydrography

Alberta Township Survey System

120°W   110°W
60°N   60°N

ALBERTA

Area Enlarged

Edmonton

Calgary

60°N

110°W

City of San Marcos General Plan

City of San Marcos

San Marcos, California, USA

By City of San Marcos and AECOM

Contact

Mettja Kuna, mkuna@san-marcos.net

Software

ArcGIS 10.1 for Desktop and Adobe Illustrator CS4

Data Sources

City of San Marcos, AECOM, SanGIS, Fehrs and Peers, Federal Emergency Management Agency, North County Transit District, US Geological Survey, US Census Bureau, California Department of Conservation, San Diego Association of Governments, California Geological Survey

The general plan is one of the most important documents for city planning. Required for all cities by the State of California, it lays out the future of the city's development in general terms through a series of policy statements depicted in text and maps. Sometimes, the general plan is referred to as a city's blueprint for achieving residents' vision for the future. City decisions related to development, growth, infrastructure, and environmental management must be consistent with the policies contained in the general plan.

In fall 2009, the City of San Marcos kicked off a comprehensive update of its general plan. Working collaboratively with AECOM, the City of San Marcos' GIS Department prepared a series of thirty-four maps to be included in the general plan. This effort was a multiyear process that required input from many city departments and various consultants. The maps range from land use to mobility improvements to future recreational uses. The City of San Marcos General Plan is a living document and will continue to be updated with changing local conditions and community priorities.

Courtesy of City of San Marcos.

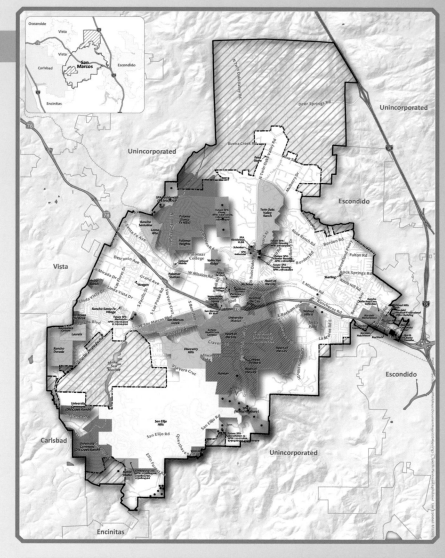

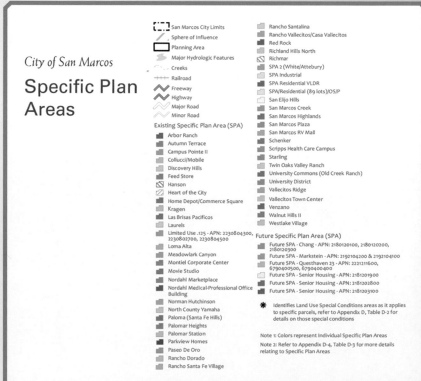

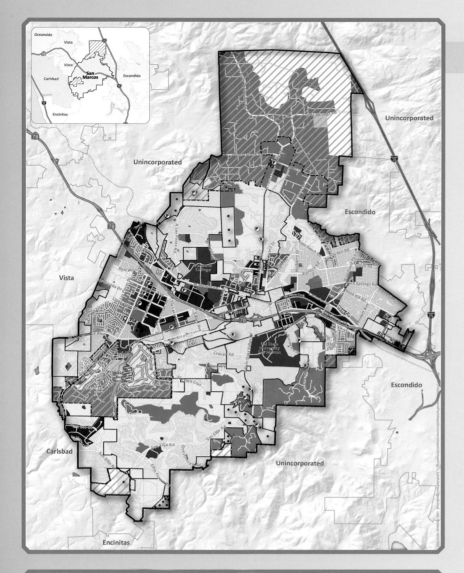

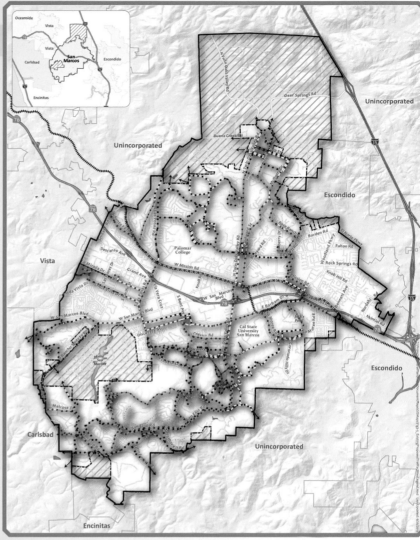

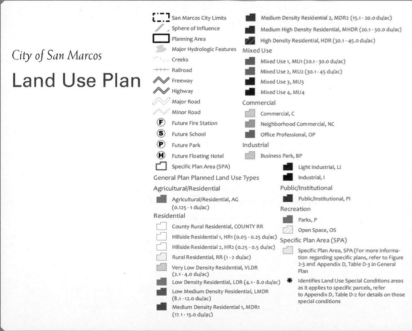

City of San Marcos

Land Use Plan

⌐⌐⌐ San Marcos City Limits
▨ Sphere of Influence
▭ Planning Area
◊ Major Hydrologic Features
〜 Creeks
┼┼┼ Railroad
∿ Freeway
∿ Highway
∿ Major Road
∿ Minor Road
Ⓕ Future Fire Station
Ⓢ Future School
Ⓟ Future Park
Ⓗ Future Floating Hotel
▢ Specific Plan Area (SPA)

General Plan Planned Land Use Types

Agricultural/Residential
▪ Agricultural/Residential, AG (0.125 - 1 du/ac)

Residential
▢ County Rural Residential, COUNTY RR
▢ Hillside Residential 1, HR1 (0.05 - 0.25 du/ac)
▢ Hillside Residential 2, HR2 (0.25 - 0.5 du/ac)
▢ Rural Residential, RR (1 - 2 du/ac)
▨ Very Low Density Residential, VLDR (2.1 - 4.0 du/ac)
▪ Low Density Residential, LDR (4.1 - 8.0 du/ac)
▪ Low Medium Density Residential, LMDR (8.1 - 12.0 du/ac)
▪ Medium Density Residential 1, MDR1 (12.1 - 15.0 du/ac)

▪ Medium Density Residential 2, MDR2 (15.1 - 20.0 du/ac)
▪ Medium High Density Residential, MHDR (20.1 - 30.0 du/ac)
▪ High Density Residential, HDR (30.1 - 45.0 du/ac)

Mixed Use
▪ Mixed Use 1, MU1 (20.1 - 30.0 du/ac)
▪ Mixed Use 2, MU2 (30.1 - 45 du/ac)
▪ Mixed Use 3, MU3
▪ Mixed Use 4, MU4

Commercial
▪ Commercial, C
▪ Neighborhood Commercial, NC
▪ Office Professional, OP

Industrial
▪ Business Park, BP
▪ Light Industrial, LI
▪ Industrial, I

Public/Institutional
▪ Public/Institutional, PI

Recreation
▪ Parks, P
▨ Open Space, OS

Specific Plan Area (SPA)
▢ Specific Plan Area, SPA (For more information regarding specific plans, refer to Figure 2-3 and Appendix D, Table D-3 in General Plan

∗ Identifies Land Use Special Conditions areas as it applies to specific parcels, refer to Appendix D, Table D-2 for details on those special conditions

City of San Marcos

Existing and Proposed Trail Facilities

⌐⌐⌐ San Marcos City Limits
▨ Sphere of Influence
▭ Planning Area
◊ Major Hydrologic Features
〜 Creeks
┼┼┼ Railroad
∿ Freeway
∿ Highway
∿ Major Road
∿ Minor Road

Trails Master Plan
∿ Multi-Use - Existing
••• Multi-Use - Future - Not On Development Plans
∿ Multi-Use - Future - On Development Plans
····· Private - Existing
•••• Private - Future - On Development Plans
〜 Sidewalk - Existing
∿ Soft Surface - Existing
••• Soft Surface - Future - Not On Development Plans
∿ Soft Surface - Future - On Development Plans
∿ Urban - Existing
•°• Urban - Future - Not On Development Plans
∿ Urban - Future - On Development Plans

→ Trails Master Plan Continuations

Estonian Basic Map

Estonian Land Board

Tallinn, Harjumaa, Estonia

By Anu Kaljula and Carmen Jaeski

Contact

Kristian Teiter, Kristian.Teiter@maaamet.ee

Software

ArcGIS 10.0 for Desktop

Data Source

Estonian Topographic Database

This project was initiated in 1991 using aerial photos, fieldwork materials, and available cartographic and statistical data as source data for mapping. At the beginning of the project, the maps were produced manually without using any computers. The digital production of the basic map started in 1996, and since 1997, the whole production process has been digital. The digital data of the basic map served as the basis for the Estonian Topographic Database (ETD), but starting in February 2009, ETD became a data source for production of basic maps.

The map shown here was one of the outputs for the Estonian Basic Map and is currently an output for the Estonian Topographic Database. In addition to ETD data, the printed map presents points of the national geodetic network and their code numbers, state border markers and their numbers, boundaries of counties and administrative units, boundaries of protected areas, immovable monuments and heritage conservation areas, isobaths, depth, and fire reservoirs.

Courtesy of MAA-AMET.

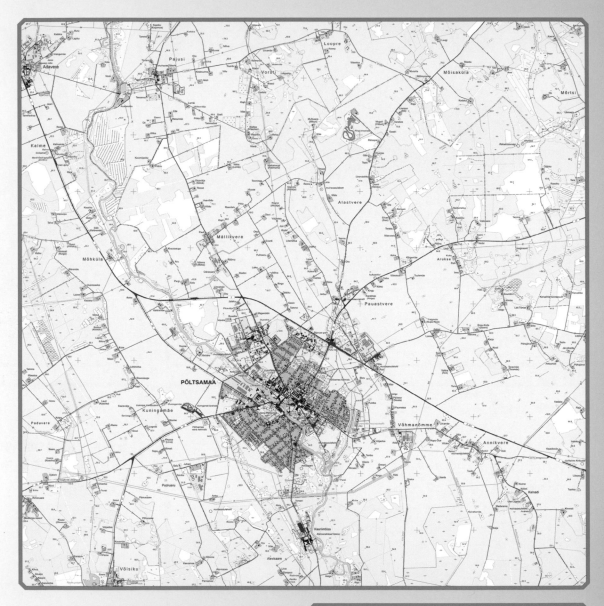

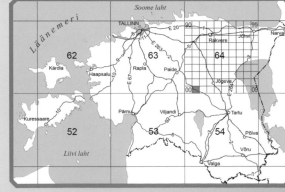

Municipality of Abu Dhabi Basemap

Municipality of Abu Dhabi City
Abu Dhabi, United Arab Emirates

By GIS support, Municipality of Abu Dhabi City

Contact

Maitha Alnuaimi, gis.Support@adm.abudhabi.ae

Software

ArcGIS 10.0 for Desktop

Data Source

Municipality of Abu Dhabi City

The sample Abu Dhabi Island basemap features the best available data from various departments of Municipality of Abu Dhabi City that includes parks, landmarks, transportation networks, buildings, beaches, and recreational facilities. It has the details of some important locations in the city and is designed to be used as a general purpose background map to support a wide variety of applications used by the municipality for various activities.

Courtesy of Municipality of Abu Dhabi City.

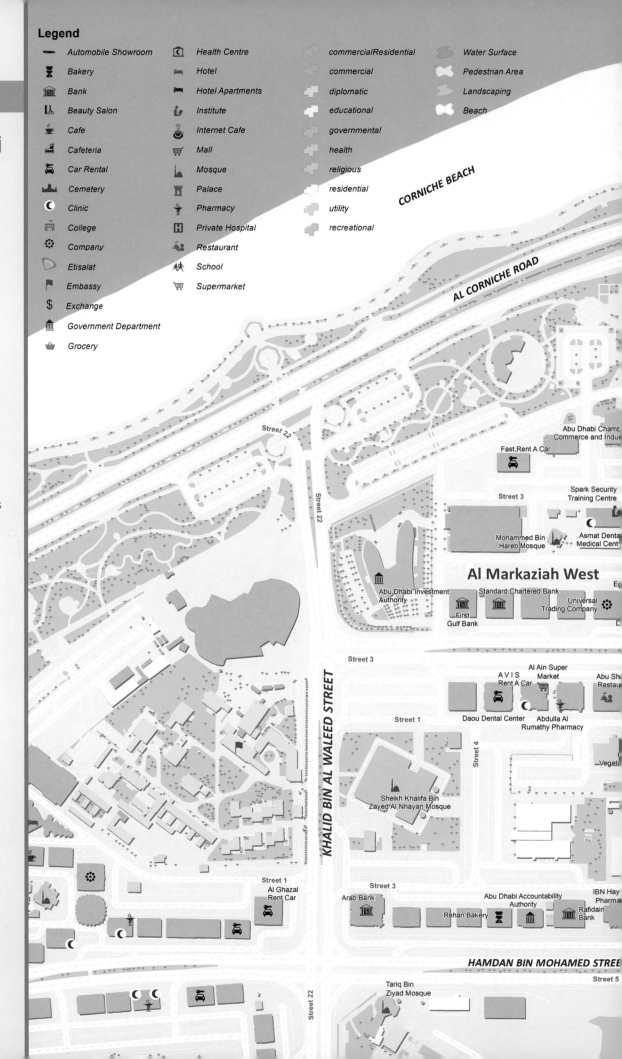

Legend

Automobile Showroom	Health Centre	commercialResidential	Water Surface
Bakery	Hotel	commercial	Pedestrian Area
Bank	Hotel Apartments	diplomatic	Landscaping
Beauty Salon	Institute	educational	Beach
Cafe	Internet Cafe	governmental	
Cafeteria	Mall	health	
Car Rental	Mosque	religious	
Cemetery	Palace	residential	
Clinic	Pharmacy	utility	
College	Private Hospital	recreational	
Company	Restaurant		
Etisalat	School		
Embassy	Supermarket		
Exchange			
Government Department			
Grocery			

Arabian Gulf

AL CORNICHE ROAD — Street 1

RASHID BIN SAEED AL MAKTOUM STREET — Street 2

FOURTH STREET

Street 2

Street 1

HDL Habib Bank

BLC Bank

First Gulf Bank

Al Manal Dental Center

National Bank of Abu Dhabi

Al Markaziyah

Najia Restaurant

Muhammed Bin Maneaa Al Remuthi Mosque

Street 1

Street 1

Sheikh Butte Bin Khadem Mosque

Emper Financial Service

Dubai Bank

ADCB Bank

Ministry of Public Work

Abu Dhabi Investment Council

Street 3

Abu Dhabi Investment Authority

Hospital Francoemirien

Commerce of Du

Al Mosque

in fe

ha Pharmacy

Street 2

Union National Bank

Al Sagr Property Management

Atiq Bin Rashid Mosque

Al Kaser Al Ainy Pharmacy

Al Sorour Bakeries

National Financial Center

Abdullah Homaid Al Hubaishi Mosque

Ali Bin Murshid Mosque

Mohammed Ahmed Yarov Mosque

International Pharmacy

Better Life Pharmacy

Al Ghazal Medical Centre

Happy Home Restaurant

Al Jawhara Dental Clinic

Street 2

Lari Financial Brokerage

Central Market

Street 2

Street 6

The Rock Restaurant

Street 11

National Hospital

aser Clinic

Eclipse Hotel Star Tower Victory Dental Medical Clinic

Mashreq Bank

City Palace Restaurant

United Arab Bank

Bainunah Pharmacy

Europcar

Laser Medical Centre

BNP Paribas

Amwaj Medical Center

Golden Fork Restaurant

Hanoi Cafe & Restaurant

Street 3

Abdaljalil Al Fahim Mosque

Street 6

KHALIFA STREET

Street 3

European Canadian Medical Centre

Fatima Al Zahra Mosque

Ideal Medical Center - LLC - Al Ain Branch

British Airways

Knights Pharmacy

Russian Eye Clinic LLC

Al Manara Pharmacy

edical

Street 6

Street 6

Street 8

Abdullah bin Mazon Mosque

Mashreq Bank

Al Noor Hospital

Dr Ishaq Zaki Al Dajani Clinic

National Bank of Abu Dhabi

Thabit Bin Amr Mosque

Street 10

Abu Dhabi Islamic Bank

Etihad Primary Health Care Centre

Dr. Ronald Mcculloch Complex

Melody Music Institute

Al Sarab Cafetaria

Dr. Ronald

National Exchange Company

$

Al Rostamani Exchange

Arabian Exchange

Ayman Dental Centre

Al Rousha Restaurant

Ahmed Amery Exchange

$

Al Bawadi Pharmacy

Food Palace Restaurant

Shamssin Sweets & Bakeries

Street 16

Street 18

Lari Exchange

$

Samiyya Ladies Salon

Southern-Wood Restaurant

Royal Beach Restaurant

Sheikh Hamid Bin Butti Mosque

$

Etisalat

$

Al Fardhan Exchange

Street 12

Sharda Vegetarian Restaurant

Al Yousif Bakeries and Sweets

Redha Al Ansari Exchange

$

Mahimat Tower Building

Bhavna Restaurant

Jamal Supermarket

Saj Jdodan

Gulf Chinese Medical Centre

Modern Pharmacy

Jamal Supermarket

Al Salah Pharmacy

Ansari Medical Centre - LCC

Golden Falcon Tower

Royal Regency Hotel

Al Salah Pharmacy

City Pharmacy Company - W L L

Bank Saderat Iran

Street 6

Ahalia Hospital

Ahalia Hospital

Holistic Dental Clinic

United Bank

Al'Qtaiba

Katia Cafe

Street 8

Al Ahlia Central Pharmacy

Haddad Medical & Dental Centre

Tarbouch Restaurant

Lulu Center

Habib Exchange

Marjan Plaza

Emirates College

Ittihad Medical Centre

Street 3

Street 4

HAMDAN BIN MOHAMED STREET

Street 5

UAE Exchange

$

Golden Tulip Delma Plaza

Wall Street Exchange

$

Al Dhafra Dental Specialized Clinic

Bank of Baroda

Al Razi Pharmacy Co.

New Land Mark Pharmacy

Central Clinic

Liwa Centre

Hotel Novotel

Street 4

Street 1

Geographic Variations in Endocrinologist Coverage for US Adults

Centers for Disease Control and Prevention

Atlanta, Georgia, USA

By Hua Lu, James B. Holt, Xingyou Zhang, Yiling J. Cheng, and Janet B. Croft

Contact

Hua Lu, hgl6@cdc.gov

Software

ArcGIS 10.1 for Desktop

Data Sources

National Provider Identifier Registry, US Census Bureau

Due to the growth and aging of the population, Americans will experience physician shortages of at least 91,500 for all specialties in 2020[1]. One such specialist is the endocrinologist, who diagnoses and treats obesity-related chronic diseases such as adult-onset diabetes, osteoporosis, thyroid disease, and metabolic syndrome.

To assess current variations in the geographic distribution of endocrinologists, the Centers for Disease Control and Prevention extracted the list of endocrinologists who serve adult patients from the National Provider Identifier Registry and geocoded their practice locations. Buffers of 5, 10, 15, 20, 30, and 50 miles surround these locations. 2010 Census Blocks and population data from the Census Bureau determine the number of US adults aged 18–64 and those aged 65 years and older who were living both within and outside each distance buffer by state, county, and urban/rural status. Coverage rates (the proportion of adults with access to at least one endocrinologist) were calculated within given distances, and the ratio of the adult population per endocrinologist.

The maps and graph here present part of the analysis results at US national, state, and county levels illustrating the current status of adult endocrinologist coverage and gaps in some geographic areas. Further research is needed to examine the impact of this variation on disease detection and control, particularly as the US population continues to age and experience obesity-related chronic diseases.

[1] Dill, M.J. and E.S. Salsberg. 2008. "The Complexities of Physician Supply and Demand: Projections Through 2025." Association of American Medical Colleges, November.

Courtesy of Centers for Disease Control and Prevention.

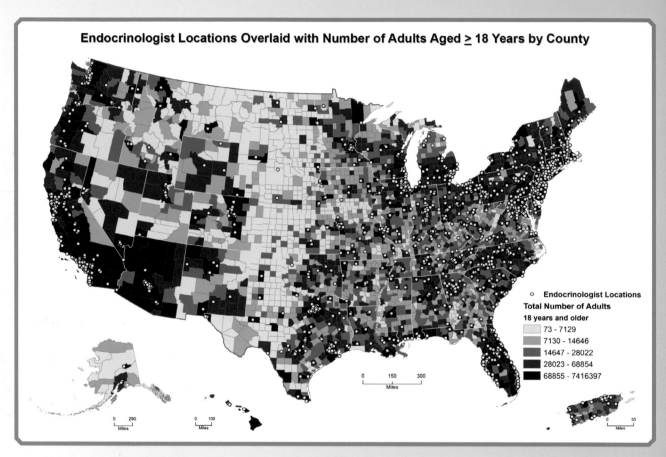

Endocrinologist Locations Overlaid with Number of Adults Aged ≥ 18 Years by County

○ Endocrinologist Locations

**Total Number of Adults
18 years and older**
- 73 - 7129
- 7130 - 14646
- 14647 - 28022
- 28023 - 68854
- 68855 - 7416397

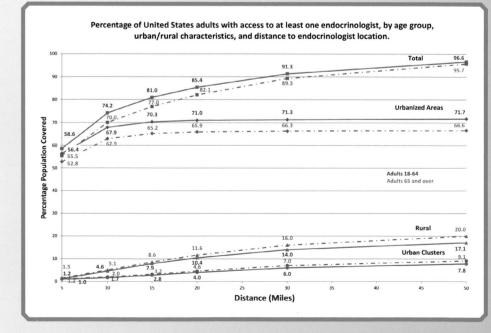

Percentage of United States adults with access to at least one endocrinologist, by age group, urban/rural characteristics, and distance to endocrinologist location.

Percentage of Adults Aged 18-64 Years Covered by at Least one Endocrinologist by County

20 miles buffer

County with
☐ 100 % Coverage

Percent
<= 10.00
10.01 - 20.00
20.01 - 30.00
30.01 - 40.00
40.01 - 50.00
50.01 - 60.00
60.01 - 70.00
70.01 - 80.00
80.01 - 90.00
90.01 - 99.99
100.00

No Coverage

Percentage of Adults Aged ≥ 65 Years Covered by at Least one Endocrinologist by County

20 miles buffer

County with
☐ 100 % Coverage

Percent
<= 10.00
10.01 - 20.00
20.01 - 30.00
30.01 - 40.00
40.01 - 50.00
50.01 - 60.00
60.01 - 70.00
70.01 - 80.00
80.01 - 90.00
90.01 - 99.99
100.00

No Coverage

Number of Covered Adults Aged 18-64 Years per Endocrinologist by County

20 miles buffer

County with
☐ 100 % Coverage

Number of Adults
(18-64 years old)
1,571 - 17,722
17,723 - 23,019
23,020 - 27,844
27,845 - 33,171
33,172 - 40,242
40,243 - 46,398
46,399 - 55,096
55,097 - 69,935
69,936 - 89,911
89,912 - 130,776
130,777 - 1,404,405

Weighted by total adults18-64 years within
distance buffer to endocrinologist

No Coverage

Number of Covered Adults Aged ≥ 65 Years per Endocrinologist by County

20 miles buffer

County with
☐ 100 % Coverage

Number of Adults
(>= 65 years old)
322 - 3,538
3,539 - 4,882
4,883 - 5,962
5,963 - 7,344
7,345 - 8,889
8,890 - 10,310
10,311 - 12,190
12,191 - 15,440
15,441 - 19,545
19,546 - 26,413
26,414 - 237,821

Weighted by total adults >= 65 years within
distance buffer to endocrinologist

No Coverage

59

Small Area Estimates of Childhood Obesity Prevalence, by Block Group, United States, 2010

Centers for Disease Control and Prevention

Atlanta, Georgia USA

By James B. Holt, Xingyou Zhang, and Stephen Onufrak

Contact

James B. Holt, jgh4@cdc.gov

Software

ArcGIS 10.0 for Desktop

Data Source

National Survey of Children's Health

Geographically detailed estimates of health burdens are important for informing public health policy and for targeting scarce public health prevention resources. However, such geographically detailed data generally is unavailable in the United States. Several ongoing public health surveillance systems collect information about important health risk factors and outcomes, chiefly through probability sample surveys, yet this data is designed to produce either national or state-level estimates. These surveys are unsuitable for generating direct survey estimates of risk factor and disease prevalence for substate areas.

Epidemiologists and researchers, therefore, must rely on statistical techniques to produce modeled estimates for small areas. In this study, the Centers for Disease Control and Prevention used a geocoded national health survey and multilevel modeling, in conjunction with a geographic information system, to illustrate the potential for estimating and mapping the prevalence of childhood obesity at the census block group level. The childhood obesity data is model based rather than direct survey or observed data. The estimates of childhood obesity prevalence were generated according to the statistically expected effects of children's age; gender; race/ethnicity; and census block group poverty, lifestyle, and urbanization on childhood obesity status. Thus, the census block group childhood obesity prevalence estimates should not be treated as the reality on the ground.

Zhang, X., S. Onufrak, J.B. Holt, and J.B. Croft. 2013. "A Multilevel Approach to Estimating Small Area Childhood Obesity Prevalence at the Census Block-Group Level." Prev Chronic Dis 10:120252. DOI: http://dx.doi.org/10.5888/pcd10.120252

Courtesy of Centers for Disease Control and Prevention.

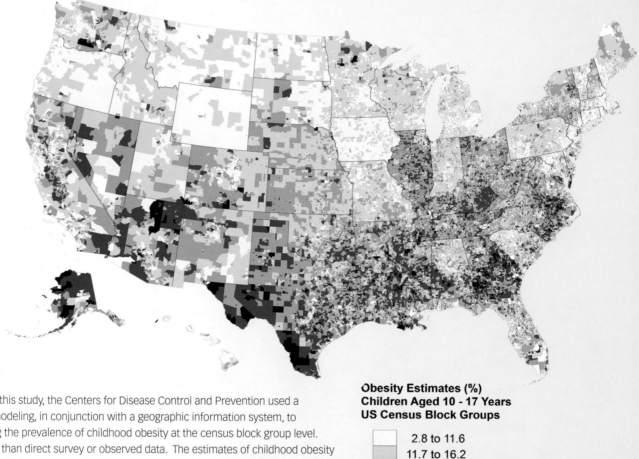

Obesity Estimates (%)
Children Aged 10 - 17 Years
US Census Block Groups

- 2.8 to 11.6
- 11.7 to 16.2
- 16.3 to 21.2
- 21.3 to 27.5
- 27.6 to 49.0
- No population aged 10-17 years

Potential Future Spatial Shifts of the Corn Belt: Influences of Climate Change

Southern Illinois University Carbondale

Carbondale, Illinois, USA

By Timothy Stoebner

Contact

Timothy Stoebner, tstoebner@siu.edu

Software

ArcGIS 10.0 for Desktop, Adobe Illustrator, SAS

Data Sources

Cropland Data Layer, US Department of Agriculture; Soil Survey Geographic database, Natural Resources Conservation Service; Global Summary of the Day, National Climatic Data Center

The current US agricultural economic environment, with its increasing demand for biofuels, food products, and exports, is influencing the need to expand and intensify crop production. The intensification of production, along with the potential for future climatic changes, could alter the landscape of American agriculture over the next century. These changes could have dramatic environmental impacts including loss of ecosystem services, decreases in water quality, and loss of natural habitats. The potential impacts are prompting a need for mitigation and adaptation through appropriate policies and management. Understanding the composition of potential future landscapes is vital in making informed decisions that could ultimately prevent unsustainable agricultural production.

The goal of this project is to analyze how future climate change and economic pressures may influence changes in specific land covers and the implications of these changes for the twenty-first century. How these land covers change over time will determine the intensity and extent of many landscapes such as the Corn Belt. Preliminary results from the corn models are represented here by corn probability maps that indicate a potential for a dramatic shift of the Corn Belt. Several future scenarios are presented to indicate a range of climate change influences depending on the level of mitigation as denoted by the three representative concentration pathways (RCPs).

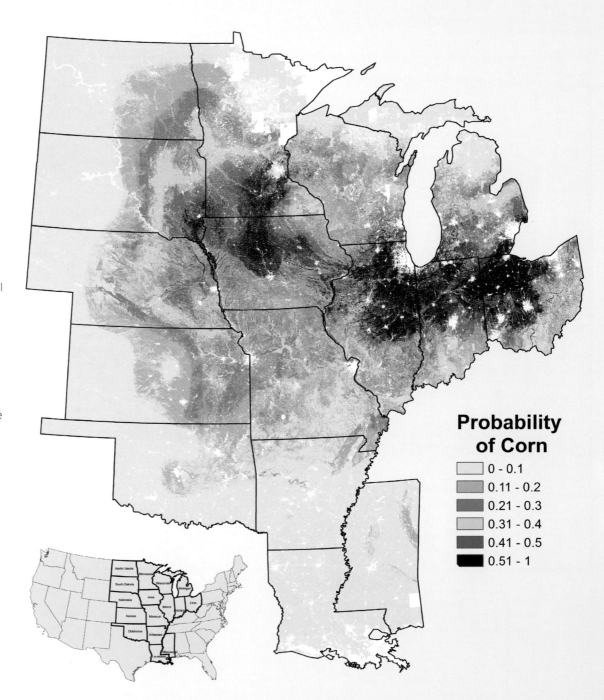

Probability of Corn

- 0 – 0.1
- 0.11 – 0.2
- 0.21 – 0.3
- 0.31 – 0.4
- 0.41 – 0.5
- 0.51 – 1

Consecutive Years of Corn Grown in the US Corn Belt (2008–2012)

US Department of Agriculture (USDA) National Agricultural Statistics Service

Fairfax, Virginia, USA

By Lee Ebinger, Dave Johnson, Patrick Willis, Claire Boryan, Audra Zakzeski, Bob Seffrin, and Karla Koudelka

Contact
Lee Ebinger, Lee.Ebinger@nass.usda.gov

Software
ArcGIS 10.0 for Desktop, ERDAS IMAGINE

Data Source
National Agricultural Statistics Service Cropland Data Layer, for years 2008–2012

Duration of Continuous Corn Plantings

- No Consecutive Years
- 2 Consecutive Years
- 3 Consecutive Years
- 4 Consecutive Years
- 5 Consecutive Years
- Cultivated Land

Average U.S. Prices by Year

- Corn $/Bushel
- Soybean $/Bushel
- Wheat $/Bushel
- Corn Ethanol $/Gallon
- Nitrogen Fertilizers $/Pound

Sources: USDA ERS, FAS, NASS, WAOB, Nebraska Energy Office

Crop rotation is a common farming practice used to manage soil nutrients, such as nitrogen, by planting different crops in consecutive years on the same land. In the United States, corn is usually grown in two-year rotations (corn-soybeans) or three-year rotations (corn-soybeans-wheat). This map shows fields in the US Corn Belt region exhibiting typical and atypical patterns of corn rotation—corn grown in nonconsecutive years and in consecutive years, respectively. Geoprocessing to determine consecutive years of corn involved extracting corn pixels from five years of Cropland Data Layer imagery, recoding the corn pixels, and then summing the five years of recoded corn imagery to produce a single image. Specific pixel values represent corn grown for two to five consecutive years or corn grown in nonconsecutive years. Corn grown for two, three, and four consecutive years occurs anytime during the five-year period from 2008 to 2012.

Courtesy of US Department of Agriculture, National Agricultural Statistics Service.

GIS-Based Estimation of Climate Change Effects on Successional Shifts in Forest Composition

US Department of Agriculture (USDA) Forest Service

Placerville, California, USA

By Arnaldo Ferreira, Debra Tatman, and Becky Estes

Contact

Debra Tatman, dtatman@fs.fed.us

Software

ArcGIS 10.0 for Desktop, Microsoft PowerPoint 2010

Data Sources

USDA Forest Service, Oregon State University, Moscow Forestry Sciences Laboratory, University of British Columbia

An exploratory study identified potential areas of the Eldorado National Forest (central Sierra Nevada mountain range) sensitive to shifts of major forest types under climate change projections for the year 2100. GIS-based models for climate and vegetation distribution were used to predict areas of potential successional changes in forest composition across the landscape. A better understanding of how climate change could alter the distribution and the composition of forests would help to adapt silvicultural practices, as well as resource allocation by the Forest Service. This study could contribute to the design of practices to cope with the projected effects of climate change on germplasm conservation. It could also be a tool in developing seed procurement strategies and distribution of tree species or populations to specific locations, minimizing the effects of climate change on long-term forest health.

Courtesy of USDA Forest Service.

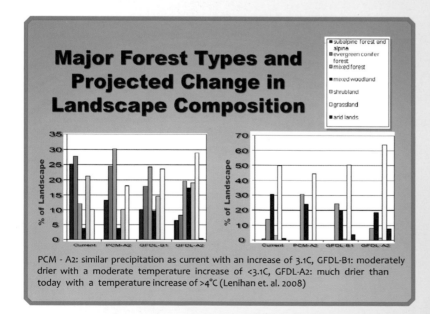

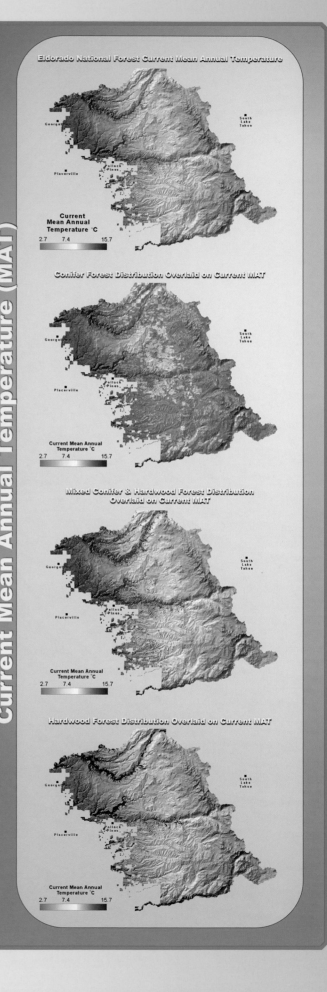

Eldorado National Forest Mean Annual Temperature Estimates for Year 2100

Future (4° Increase) Mean Annual Temperature °C
2.7   7.4   15.7+

2005 Conifer Forest Distribution Overlaid on Year 2100 MAT

Future (4° Increase) Mean Annual Temperature °C
2.7   7.4   15.7+

2005 Mixed Conifer & Hardwood Forest Distribution Overlaid on Year 2100 MAT

Future (4° Increase) Mean Annual Temperature °C
2.7   7.4   15.7+

2005 Hardwood Forest Distribution Overlaid on Year 2100 MAT

Future (4° Increase) Mean Annual Temperature °C
2.7   7.4   15.7+

Current Forest Distribution Overlaid on Suitable Land Base

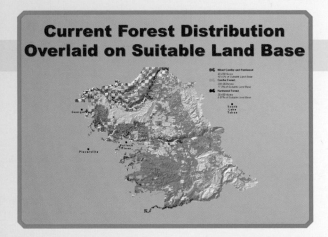

Future Forest Distribution Overlaid on Suitable Land Base

Current Forest Distribution Overlaid on Year 2100 MAT

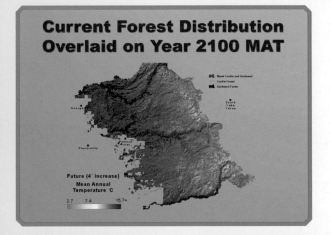

Future (4° Increase) Mean Annual Temperature °C
2.7   7.4   15.7+

Future Forest Distribution Overlaid on Year 2100 MAT

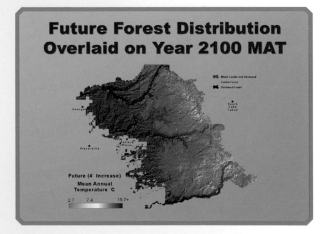

Future (4° Increase) Mean Annual Temperature °C
2.7   7.4   15.7+

65

Predominant Forest Carbon Pools of the Contiguous United States (Circa 2005)

US Department of Agriculture (USDA) Forest Service, Northern Research Station

Newtown Square, Pennsylvania, USA

By Barry Tyler Wilson, Christopher W. Woodall, and Douglas M. Griffith

Contact

Douglas Griffith, dgriffith@fs.fed.us

Software

ArcGIS Desktop 9.2

Data Sources

USDA Forest Service, Forest Inventory and Analysis Program; US Geological Survey; Esri; National Aeronautics and Space Administration

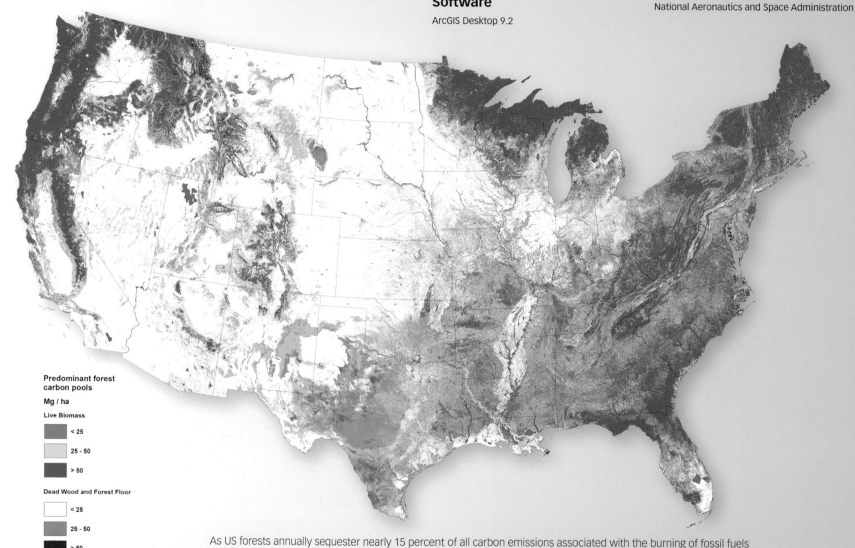

Predominant forest carbon pools

Mg / ha

Live Biomass

- < 25
- 25 - 50
- > 50

Dead Wood and Forest Floor

- < 25
- 25 - 50
- > 50

Soil Organic

- < 25
- 25 - 50
- > 50
- no estimate
- water

As US forests annually sequester nearly 15 percent of all carbon emissions associated with the burning of fossil fuels across the country, the assessment of their size, condition, and location is paramount to forest management strategies and US climate change policies. Estimates of forest carbon density were developed for the conterminous United States using the Forest Service's annual forest inventory.

Carbon density was mapped according to one of three forest carbon pools (live biomass, dead wood and forest floor, and soils) that accounted for the majority/plurality of total forest carbon within each pixel. The carbon density of pools closely related to detrital input, such as deadwood, is often highest in forests suffering from recent mortality events such as beetle infestation in the northern Rocky Mountains. In contrast, live tree carbon density is often highest on the highest-quality forest sites such as those found in the Pacific Northwest. Validation results suggest strong agreement between the estimates produced from the forest inventory plots and those from the imputed maps.

Forest inventory imputed plot maps provide an efficient and flexible approach to monitoring diverse carbon pools at national and regional scales while also allowing timely incorporation of empirical data that empowers the assessment of future climate change-related events.

Courtesy of USDA Forest Service.

Trinity Ridge Long-Term Assessment

US Department of Agriculture (USDA) Forest Service

Auburn, California, USA

By Trevor Miller, Joseph Kafka, and Tate Fischer

Contact

Trevor Miller, trevormiller@fs.fed.us

Software

ArcGIS 10.0 for Desktop

Data Sources

Wildland Fire Detection Support System, LANDFIRE, Boise National Forest data layers, National Oceanic and Atmospheric Administration-National Weather Service, Eastern Great Basin Predictive Services

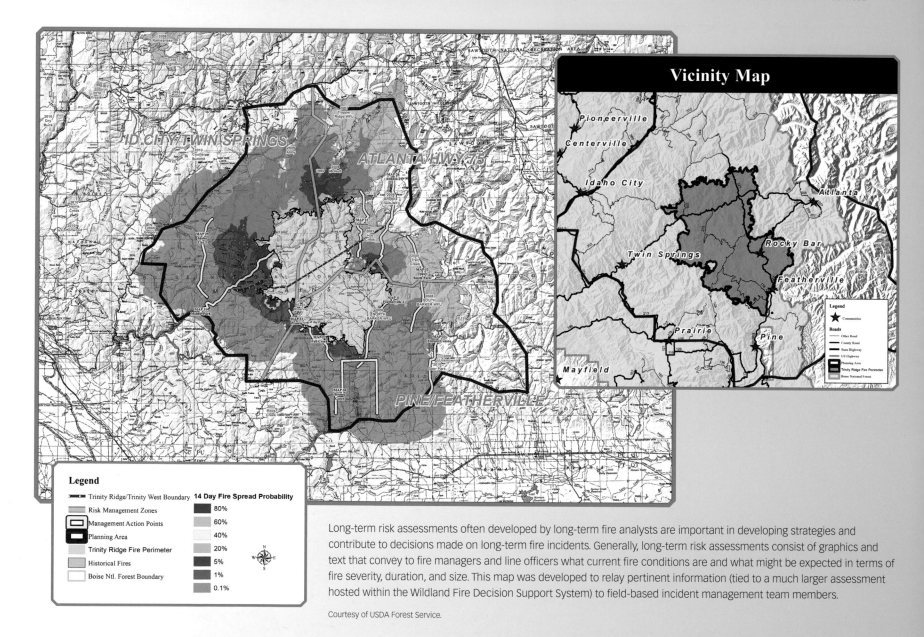

Long-term risk assessments often developed by long-term fire analysts are important in developing strategies and contribute to decisions made on long-term fire incidents. Generally, long-term risk assessments consist of graphics and text that convey to fire managers and line officers what current fire conditions are and what might be expected in terms of fire severity, duration, and size. This map was developed to relay pertinent information (tied to a much larger assessment hosted within the Wildland Fire Decision Support System) to field-based incident management team members.

Courtesy of USDA Forest Service.

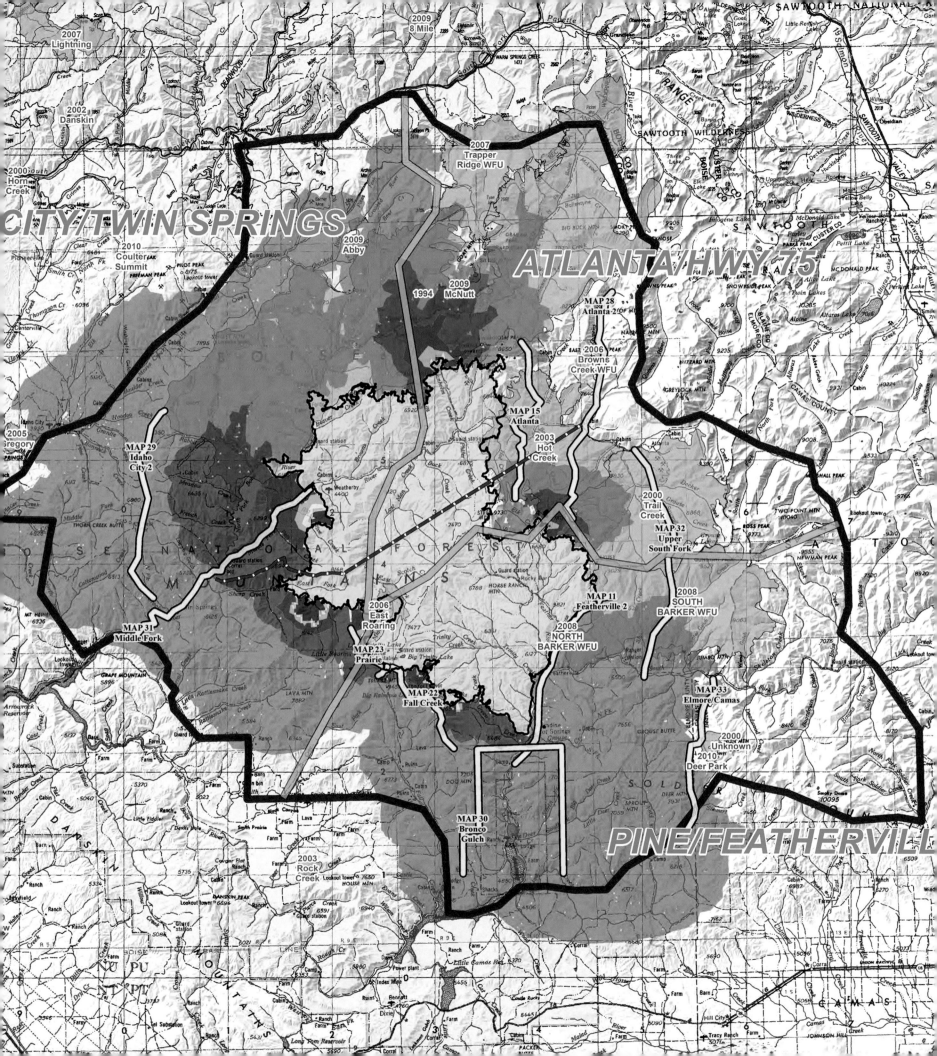

Wildland Fire Potential for the Conterminous United States

Data Sources

Fire Program Analysis system burn probability and conditional flame length probability rasters; LANDFIRE 2008 raster layers, including Fire Behavior Fuel Model 40, Existing Vegetation Type, and forest canopy characteristics.

US Department of Agriculture (USDA) Forest Service

Missoula, Montana, USA

By Gregory K. Dillon

Contact

Gregory K. Dillon, gdillon@fs.fed.us

Software

ArcGIS 9.3 and 10.0 for Desktop

Legend

National Forest Administrative Boundaries

Wildland Fire Potential

Very Low

Low

Moderate

High

Very High

Developed Lands

Non-burnable Lands*

Water

The Wildland Fire Potential (WFP) map is a raster geospatial product produced by the Forest Service Fire Modeling Institute for use in analyses of wildfire risk or hazardous fuels prioritization at large landscapes (hundreds of square miles) up through regional or national scales. The WFP map builds on and integrates estimates of burn probability and conditional probabilities of fire intensity levels generated for the national interagency Fire Program Analysis (FPA) system using a simulation modeling system called the Large Fire Simulator (FSim). The specific objective of the 2012 WFP map is to depict the relative potential for wildfire that would be difficult for suppression resources to contain, based on past fire occurrence, 2008 fuels data from LANDFIRE, and estimates of wildfire likelihood and intensity generated from FSim in 2012. Areas with higher WFP values, therefore, represent fuels with a higher probability of experiencing high-intensity fire with torching, crowning, and other forms of extreme fire behavior under fire-conducive weather conditions.

On its own, WFP does not provide an explicit map of wildfire threat or risk because no information on the effects of wildfire on specific resources and assets such as habitats, or infrastructure is incorporated in its development. However, the WFP map could be used to create value-specific risk maps when paired with spatial data depicting highly valued resources. WFP is also not a forecast or wildfire outlook for any particular season, as it does not include any information on current or forecast weather or fuel moisture conditions. It is instead intended for long-term strategic planning and fuels management.

Courtesy of USDA Forest Service.

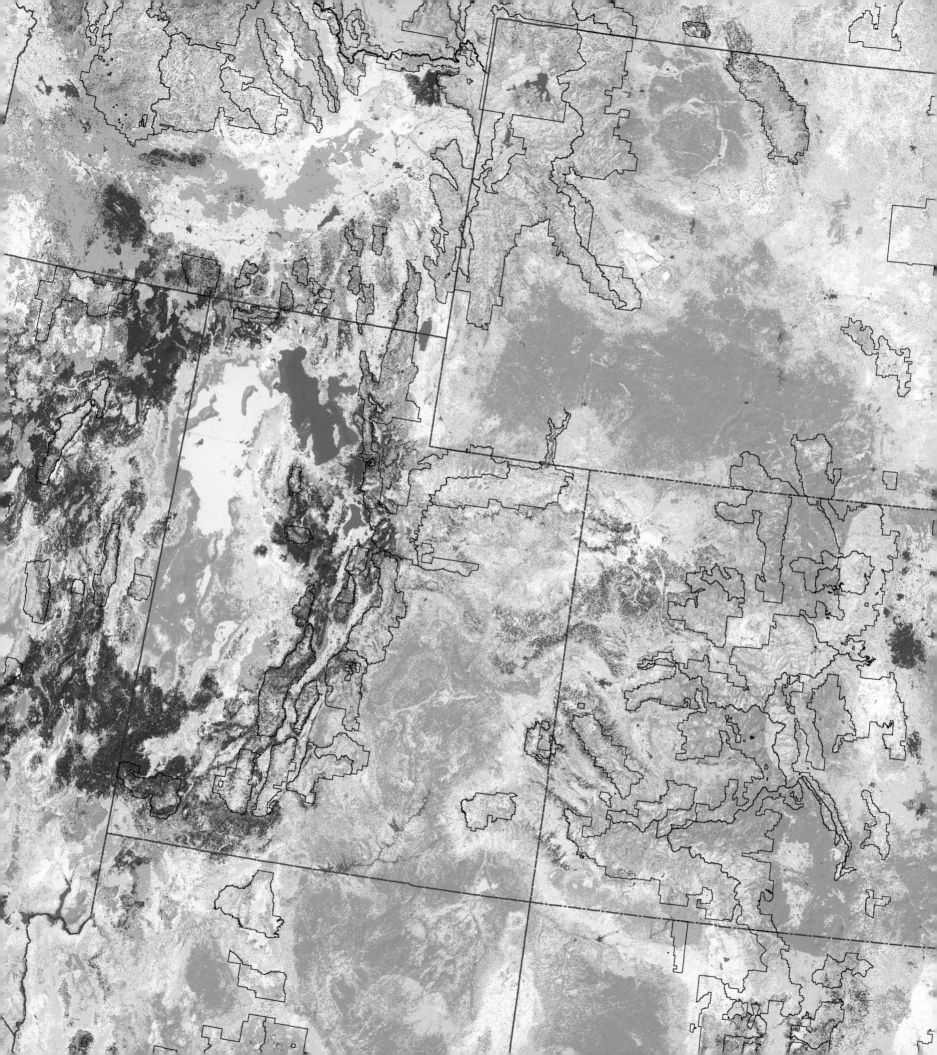

Large Wildfire Probability Modeling—Oregon

US Department of Agriculture (USDA) Forest Service

Corvallis, Oregon, USA

By Cole Belongie, Jeremy Hobson, and Ray Davis

Contact

Cole Belongie, cbelongie@fs.fed.us

Software

ArcGIS 10.1 for Desktop, MaxEnt ver. 3.3.3k

Data Sources

USDA Forest Service, Oregon Department of Forestry, Oregon State University, US Gelogical Survey

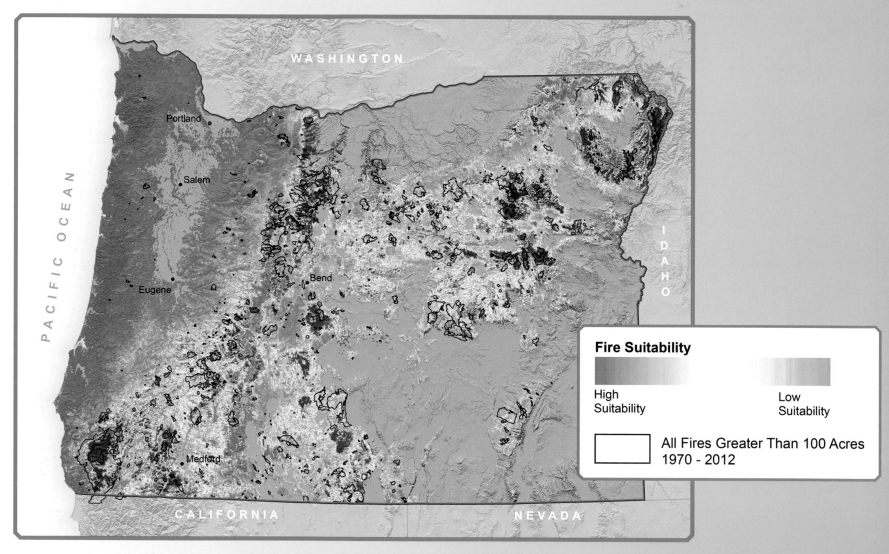

Understanding the spatial patterns of past, present, and future wildfire environments is key for forest management organizations to devise sound strategies for fuels and fire management. In Oregon, wildfires pose challenges to managers involved with protecting critical habitats as well as preventing catastrophic fires in the wildland-urban interfaces. A collaborative project to model the probability of large wildfire occurrence across forested lands of Oregon was undertaken between the Oregon Department of Forestry, Oregon State University, and the US Department of Agriculture (USDA) Forest Service, Region 6.

MaxEnt (Maximum Entropy) was one of three statistical modeling techniques used to produce maps of Oregon's forests that show the entire gradient of environmental conditions relating to threats of large wildfires. A final map displays the predicted likelihood of large wildfire occurrence, which varies geographically. Results from the 2013 fire season showed positive correlations between new fires and areas predicted as having environments highly suitable for the occurrence of large wildfires.

Courtesy of USDA Forest Service.

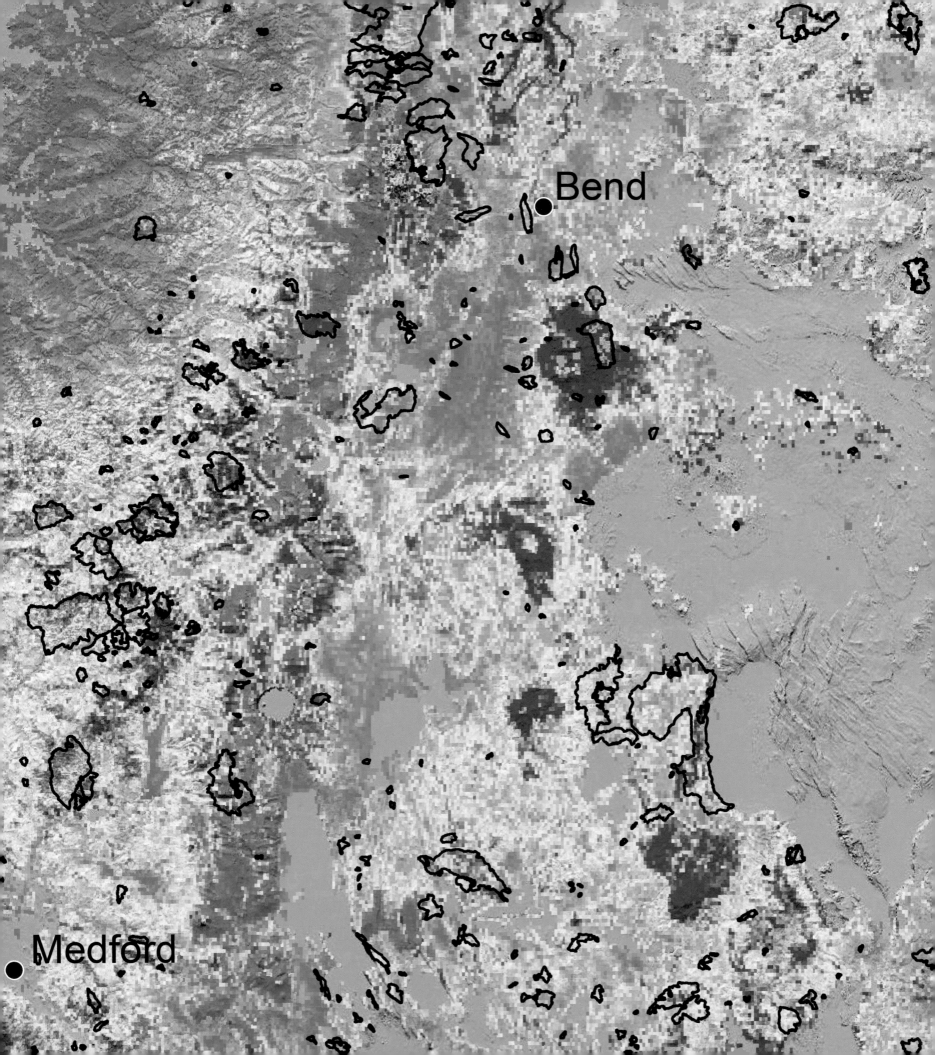

Mapping Earthen Berms in the Meadowlands

New Jersey Meadowlands Commission

Lyndhurst, New Jersey, USA

By Francisco Artigas, Dom Elefante, Sal Kojak, Adam Osborn, Stephanie Bosits, and Ildiko Pechmann

Contact

Dom Elefante, Dominador.Elefante@njmeadowlands.gov

Software

ArcGIS 10.0 for Desktop, Arc Hydro, ArcGIS Spatial Analyst, ArcGIS 3D Analyst, ENVI 5.0, Quick Terrain Modeler 8.02

Data Source

Meadowlands Environmental Research Institute

In October 2012, Hurricane Sandy moved up the Eastern Seaboard, merged with two other substantial low-pressure systems, and came ashore during a spring high tide. Landfall of the resultant 1,000-mile-wide storm was centered on the New Jersey coastline. In the Hackensack River estuary, the towns of Carlstadt, Little Ferry, and Moonachie experienced unprecedented flooding. As a result, residential and business areas experienced floodwater levels that reached 2 to 3 feet above street level. The Meadowlands Environmental Research Institute (MERI), the scientific arm of the New Jersey Meadowlands Commission (NJMC), continuously monitors the Hackensack River, and at midnight on October 29, 2012, water level sensors in the vicinity of Carlstadt and Moonachie recorded depths of 8.5 to 9.5 feet (NADV88) that lasted six hours above 7 feet. Water levels of this height and duration overtopped all legacy earthen berms installed by the Mosquito Commission between 1913 and the early 1970s. The elevation of these berms averages 5 feet above sea level, and they were not originally intended as flood control structures but as part of a mosquito control strategy and designed to prevent standing water on the land side of the berms. There are hundreds of miles of ditches and berms in this area that have not been maintained since the late 1970s.

To assist the towns, MERI-GIS focused on identifying vulnerable areas (soft edges) where the existing berms are significantly lower than the original 5 feet. These soft edges are considered potential weak points to future floods, and therefore, efforts were made to identify and map these vulnerable areas. These maps show the steps that were followed to detect soft edges within the affected area. The results from this study are being used to map vulnerable areas within the Meadowlands coastal communities.

Courtesy of New Jersey Meadowlands Commission, Meadowlands Environmental Research Institute.

Arctic Ocean Wall Map

The Pew Charitable Trusts—International Arctic Program

Washington, DC, USA

By Jeremy Davies

Contact

Jeremy Davies, jdavies@pewtrusts.org

Software

ArcGIS 10.1 for Desktop, Adobe Illustrator CS3

Data Sources

IBCAO v3, Natural Earth, GEONet Names Server, IBRU

The Arctic Ocean is one of the planet's pristine marine regions. But permanent ice is diminishing due to climate change, opening the international waters of the Central Arctic Ocean to commercial fishing for the first time in human history. These waters, encompassing an area as big as the Mediterranean Sea or one-third the size of the continental United States, are not governed by a fisheries agreement. Such an accord is needed to close this region to commercial fishing until scientific knowledge and management measures can ensure a sustainable fishery.

The Pew Charitable Trusts' international Arctic campaign produced this wall map to serve a variety of purposes:

1. To create an up-to-date map of the Arctic Ocean and surrounding areas that includes comprehensive labeling of surface and subsurface marine features as well as important land features. It serves as a reference map for anyone with an interest in the high Arctic.

2. To delineate clearly the 200-nautical-mile maritime boundary within which the international waters of the Central Arctic Ocean are vulnerable to the start of unregulated commercial fishing unless an international fisheries agreement is negotiated.

3. To accurately represent the complex and interesting seafloor relief of the Arctic Ocean, which is increasingly relevant as multiyear ice is replaced by open water and seasonal ice. Marine topography is one of the main drivers of fisheries productivity.

Courtesy of The Pew Charitable Trusts, 2013.

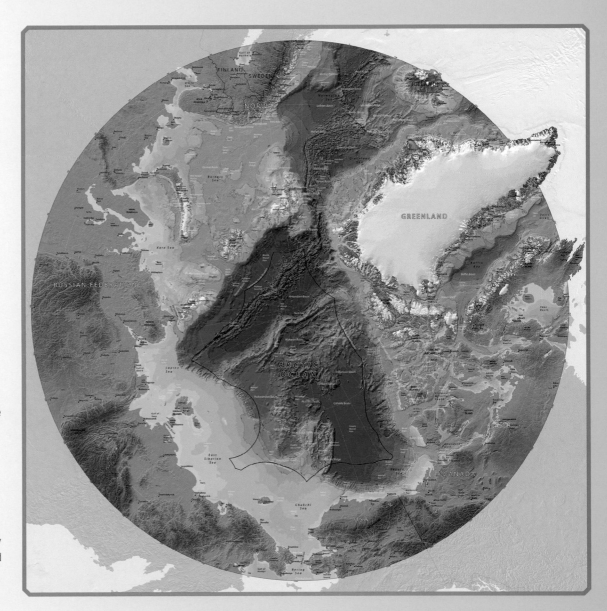

Novyy
Port
Yamburg
ovskiy
...odka
GYDA
PENINSULA
Gyda
GYDA
PENINSULA
Belyy I.
YAMAL
PENINSULA
Gulf
of Ob
Yenisey Gulf
Dickson
Vorontsovo
Dudinka
Norilsk
Agapa
Pyasina
Kara Sea
Arctic
Institute I.
Mikhaylova
Central Kara Plateau
Vize I.
Ushakova I.
Voronin
Trough
FEDERATION
Boyarka
Shmidta I.
October
Revolution I.
Komsomolets I.
Khatanga
Khatanga
SEVERNAYA
ZEMLYA
Nizhnyaya
Taymyra
Staroybnoye
Vil'kitskogo
Strait
Bol'shevik I.
TAYMYR
PENINSULA
Chelyuskin
Cape
Chelyuskin
Khatanga
Gulf
Nordvik
Bolshoy
Begichev I.
Khorgo
Bol'shaya
Kuonamka
Saskylakh
Anabar
Zhilinda
Olenek
Ust
Olensk
Gulf of
Olen'k
Lena
Delta
Sagastyr
Tumatskaya
Protoka
Trofimovsk
...kere
Lena
Govorovo
Natara
Udzha
Tiksi
Olenek
Laptev
Sea
Olenek
Valley
Yana
Valley
Stolbovoya
Bank
Vasil'yev
Shoal
Kotel'nyy I.
...k
NEW
SIBERIAN
ISLANDS
Zemlya
Bunge
of
Lyakhov
Islands
Ust
Kuyga
Kavache
Shalaurova
Novaya
Sibir' I.
Dmit...
Lap...

NOVAYA
ZEMLYA
East
Novaya
Zemlya
Trough
Admiralteystva
Rise
Medvezhiy
Trough
Cape
Zhelaniya
SVALBARD
Olga Basin
Spitsbergen
Kong Karls Basin
Nordaustlandet
Kvitoya
Franz Viktoriya
Trough
FRANZ
JOSEF
LAND
Zemlya
Georga I.
Greem
Bell I.
Admiralteystva
Rise
Svyataya
Anna Trough
North
Kanin
Bank
Barents
Abyssal
Plain
Central Arctic Ocean Boundary
Kucherov
Terrace
Voronov
Terrace
Nansen Basin
Svyataya
Anna Fan
Gakkel Ridge
Persey
Rise
...b
Bank
Hinl...
Trou...
Pole
Abyssal
Plain
An...
Rassokho
Seamounts
Lomonosov Ridge
ARLIS
Gap
AR
OC
Makar...
Oden
Spur
Sib...
Aby...
Pla...
Geofizikov
Spur
Wrangel
Plain
Podvodnikov Basin
Mendeleyev
Ridge

∧ **Central Arctic Ocean Boundary**

Depth in Metres

-5500 -5000 -4500 -4000 -3500 -3000 -2500 -2000 -1500 -1000 -500 -200 -100 -50 -25 -10

GREENLAND

Qasigiannguit
Ilulissat
Disko
Bay
Uummannaq
Qeqertasuaq
Uummannaq
Qeqertasuaq
Disko
Bank

Spitsbergen
Fracture
Zone

AWI Bank

Ob' Bank

Belgica Bank

Northwind
Shoal

nak Plateau

drenna

jubre
ank

pet

Kangersuatsiaq
Upernavik
Tasiusaq

Nord

Nuussuaq
Kullorsuaq

Lena
Trough

Morris
Jesup
Rise

Melville
Bay

Baffin
Bay

Clyde
River

Savissivik

Baffin Basin

Lincoln
Sea

Devon
Slope

Robeson
Channel
Hall
Basin

Alert

Kennedy
Channel

Qaanaaq

Murch...

Kane
Basin

North
Water
Polynya

Talbot
Trough

Devon Shelf

Basin

ELLESMERE
ISLAND

Pond
Inlet

Bylot
Island

Grise
Fjord

Eureka

Jones
Sound

Lancaster
Sound

Lancaster
Trough

Nanisivik
Arctic Bay

Axel
Heiberg I.

QUEEN
ELIZABETH
ISLANDS

Devon
Island

Marvin
Spur

Alpha Ridge

Alpha
Cordillera

Central Arctic Ocean Boundary

Queen
Elizabeth
Rise

Barrow Sill

Queen Elizabeth Shelf

Gulf of
Booth

ARCTIC
ARCHIPELAGO

Cornwallis
Resolute
Somerset I.

TIC
AN

Stefansson Basin

Bathurst I.

Parry
Channel

Boothia
Peninsula

Taloyc

Prince of
Wales I.

Larsen
Sound

King
William I.

Gustaf
Adolf
Trough

Lougheed
Basin

Lamont Ridge

M'Clintock
Channel

Goldsmith
Channel

PARRY
ISLANDS

Prince
Patrick I.

Melville
Island

Melville
Trough

Viscount
Melville
Sound

Hadley Bay

Canada Basin

utilis Basin

Nautilus
Basin

inabar-Khatanga
Valley

Richard
Collinson
Inlet

M'Clure
Strait

VICTORIA
ISLAND

Cambridge
Bay

Umingm

Bathurst
BANKS

Geology of the Alberta Rocky Mountains and Foothills

Alberta Energy Regulator— Alberta Geological Survey

Edmonton, Alberta, Canada

By Dinu Pana and Rastislav Elgr

Contact

Rastislav Elgr, rastislav.elgr@aer.ca

Software

ArcGIS Desktop 9.3.1, InDesign CS5, Canvas 12, GIMP 2.8

Data Source

Alberta Geological Survey

The Alberta Geological Survey (AGS) has produced the first stand-alone, updated, and seamless geological compilation map of the Alberta portion of the Rocky Mountains and Foothills. The scale of 1:500,000 was selected to display the entire region at sufficient detail and in a manageable size when printed.

The Alberta Rockies and Foothills are part of the continental-scale foreland fold-and-thrust belt, which forms the eastern margin of the North American cordillera. This compilation map encompasses the Alberta portion of the foreland belt, from the Alberta–British Columbia border to the US border. Due to its high economic potential, this orogenic segment has been the focus of intense geological exploration and research for over a century and was mapped by different generations of workers at scales ranging from 1:50,000 and 1:63,360 to 1:1,000,000. Therefore, the geological information included on existing geological maps varies widely between areas with uneven levels of detail and accuracy.

The AGS compilation map incorporates a consistent level of detail for the Alberta portion of the Cordilleran orogen. This was achieved by the systematic acquisition of all source materials; filtering and restructuring of the geological information; and digitizing and georeferencing all existing hard-copy geological maps published by government agencies, academia, and industry. AGS has conducted extensive fieldwork to verify contacts on old maps and to resolve map edge discrepancies as well as various inconsistencies across boundaries between local maps with uneven geological information, inconsistent nomenclature, or structural and stratigraphic interpretations. The compilation map can be readily updated and republished with the addition of new layers as the need arises.

Courtesy of Alberta Energy Regulator—Alberta Geological Survey.

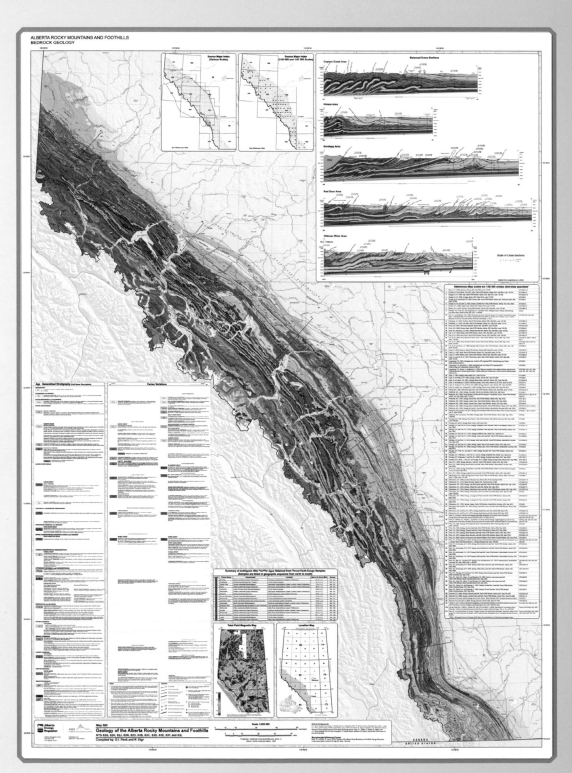

Alaska and Yukon Territory Mineral and Energy Resources

HDR Inc.
Anchorage, Alaska, USA

By Scott M. Norton

Contact

Scott Norton, scott.norton@hdrinc.com

Software

ArcGIS 10.1 for Desktop, Adobe Illustrator CS6

Data Sources

Alaska Department of Natural Resources, Alaska Oil and Gas Conservation Commission, Yukon Geological Survey, Mineral Services Unit, Alaska Department of Transportation

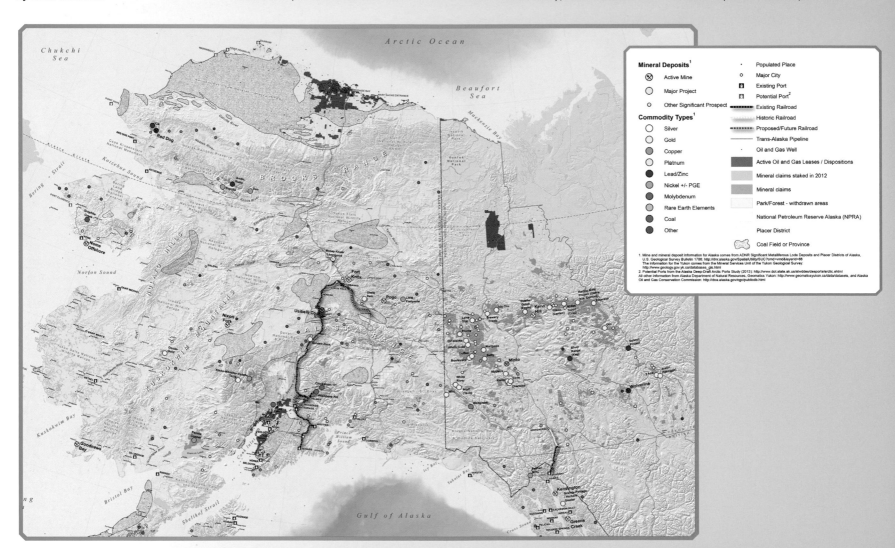

Railroads shaped Alaska's history and the development of its natural resources. Today, rail remains a key piece of Alaska's transportation infrastructure that can facilitate further regional economic activity through resource development and is essential to the development of resources such as coal.

The Alaska Department of Transportation and Public Facilities (ADOT&PF) is developing a statewide rail plan to provide direction for future policy involving freight and passenger rail transportation in the state. The Alaska State Rail Plan (ASRP) involves agency and public collaboration and will align Alaska's rail priorities based on the common interests of the stakeholders statewide.

HDR Inc. was contracted to assist ADOT&PF with the development of the ASRP, and its responsibilities include the stakeholder engagement portion of the planning process. In addition to its use as a general reference document, this map was developed for public meetings, outreach workshops, and meetings with various stakeholder groups to help spur ideas and discussion about priorities for future rail infrastructure development. The map shows the locations of significant mineral and energy resources and communities throughout Alaska and the Yukon Territory that could benefit from rail along with existing and potential port sites. The map also shows physical barriers (mountain ranges) and administrative barriers (state and federal parks and wildlife refuges) that could make rail infrastructure development more challenging.

Courtesy of HDR Inc.

Freshwater Pigment Retrieval with Satellite Images: Lake Atitlán Case Study

University of Alabama in Huntsville

Huntsville, Alabama, USA

By Africa I. Flores Cordova, Robert Griffin, Sundar Christopher, and Daniel Irwin

Contact

Africa I. Flores Cordova, africa.flores@nsstc.uah.edu

Software

ArcGIS 10.0 for Desktop, ENVI

Data Sources

Landsat 7 ETM+, Landsat 5 TM, EO-1 Ali and Hyperion

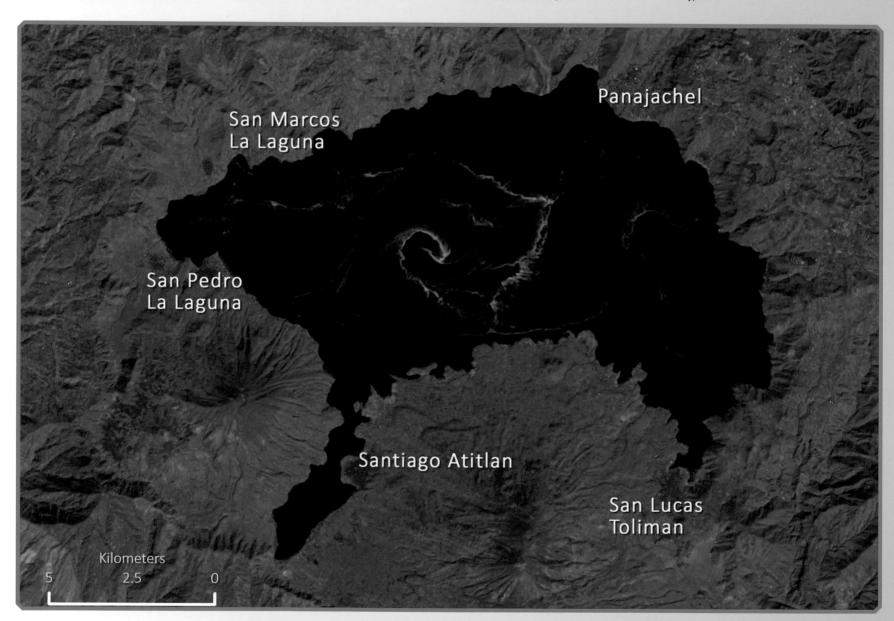

This map depicts the algal bloom that covered Lake Atitlán, located in the Guatemala highlands, in 2009. The nutrients feeding the bloom came from sewage, agricultural runoff, and increased runoff as a result of deforestation around the lake basin. The area covered by the algal bloom was estimated using a variety of remote sensors. The sequence of images in this map portrays the progress of the algal bloom in the lake from October 30 to the end of December 2009. This work was done through SERVIR, a joint project between the National Aeronautics and Space Administration (NASA) and the US Agency for International Development (USAID).

Special thanks to CATHALAC, Betzy Hernandez, and Emil Cherrington for their support in satellite image processing.

Courtesy of University of Alabama in Huntsville, NASA/USAID SERVIR.

Landsat
30 October 2009

EO-1 Hyperion
8 November 2009

EO-1 ALI
13 November 2009

Landsat
15 November 2009

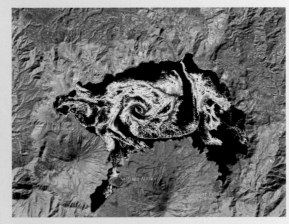

ASTER
22 November 2009

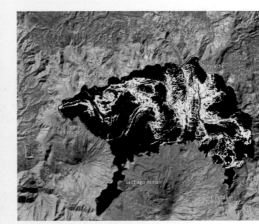

ASTER
1 December 2009

EO-1 ALI
17 December 2009

EO-1 ALI
27 December 2009

Relative algae concentration

 High

Medium

Low

Seismicity of the Earth (1900–2012): Java and Vicinity

US Geological Survey
Golden, Colorado, USA

By Eric S. Jones, Gavin Hayes, Fransiska K. Dannemann, Melissa Bernardino, Kevin Furlong, Antonio Villaseñor, and Harley Benz

Contact

Eric Jones, esjones@usgs.gov

Software

ArcGIS 10.1 for Desktop

Data Sources

Engdahl, E. R., and A. Villasenor. 2002. "Global Seismicity: 1900–1999" in *International Earthquake and Engineering Seismology Part A.* 665–690. New York: Elsevier Academic Press; and annual supplements for the interval 1900–2007, where the magnitude floor is 5.5 globally, and the NEIC PDE Catalog M4.0 and larger from 1973 to July 19, 2012

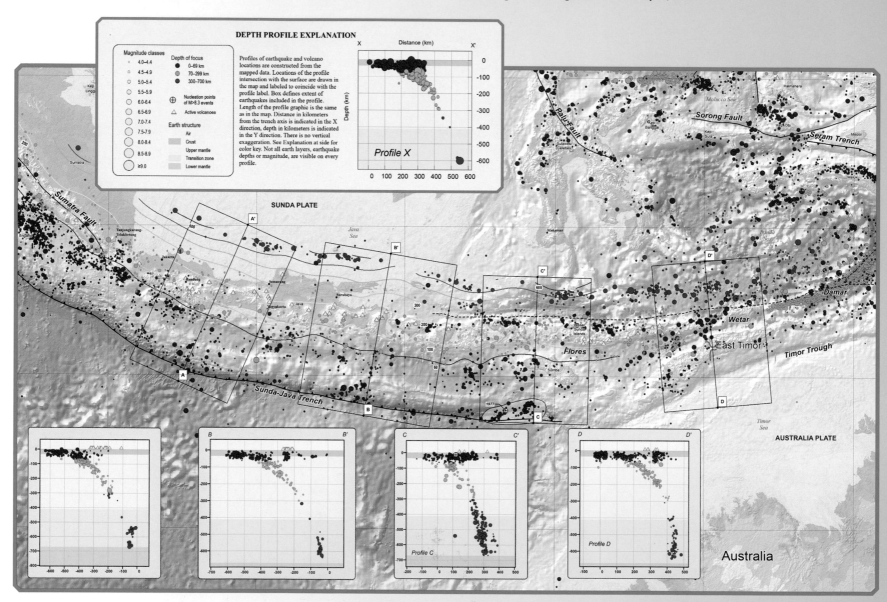

This map shows the seismicity of the earth from 1900 to 2012 in Java and its vicinity, magnitude 4.0 and greater. The main map shows four earthquake catalogs, which display a complete earthquake catalog from 1900 to 2012. The *Seismic Hazard and Relative Plate Motion* map shows the seismic hazard layer that displays the generalized seismic hazard of the region. The relative plate motion vectors show the movement of the plate according to the MORVAL model in the *Seismic Hazard and Relative Plate Motion* map as well. Depth profiles are also used to show the depth and shape of the seismicity in the Java trench.

Courtesy of US Geological Survey—Hazards.

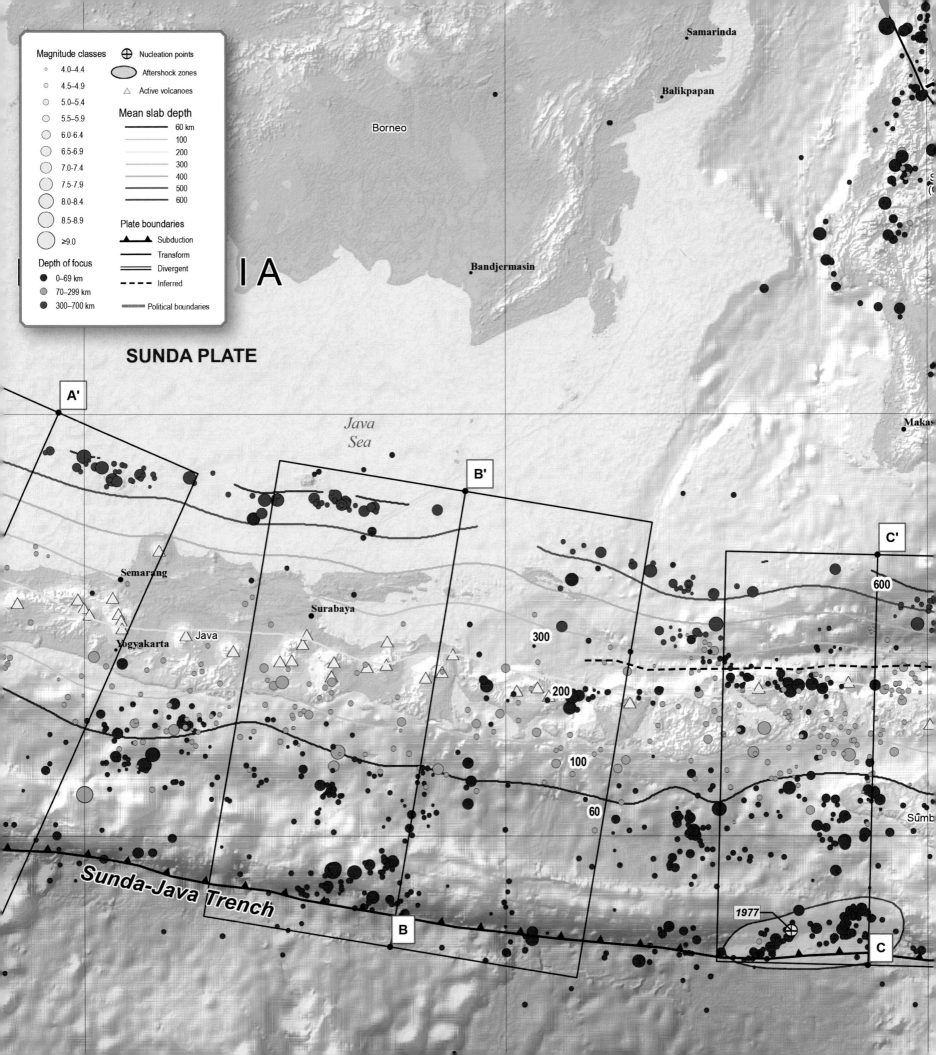

Arctic Ocean Biological Productivity Map Booklet

The Pew Charitable Trusts—International Arctic Program

Washington, DC, USA

By Jeremy Davies, Patricia Chambers, and the International Arctic Team

Contact

Jeremy Davies, jdavies@pewtrusts.org

Software

ArcGIS for Desktop, Adobe Illustrator, InDesign

Data Sources

Esri, Russian Arctic Biogeographical Database [Belikov S., Boltunov A., Belikova T., Gorbunov Yu. and Prydatko V. (2011), 1957-2011 data], Ocean Biogeographic Information System, International Bathymetric Chart of the Arctic Ocean, International Boundaries Research Unit, sailwx.info

There is a common perception that the waters of the high Arctic are a biological desert. The Pew Charitable Trusts' international Arctic campaign created a map booklet to show that although there are many unknowns about the distribution and abundance of biota in the Central Arctic

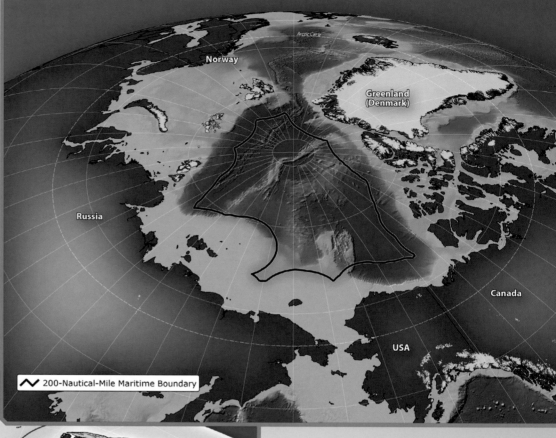

⌁ 200-Nautical-Mile Maritime Boundary

Icebreaker Tracks, 2011-2012
- **⌁** Healy (U.S.)
- **⌁** Louis St. Laurent (Canada)
- **⌁** Oden (Sweden)
- **⌁** Polarstern (Germany)
- **⌁** Xue Long- Approx. 2012 Route (China)
- **⌁** 200-Nautical-Mile Maritime Boundary

Ocean, the area is far from barren. The region is home to fish, invertebrates, migratory birds, and marine mammals. Algae grow in and beneath the sea ice, blooming in spring to fuel a food web that includes plankton, Arctic cod, ringed seals, and polar bears. Dead plankton and other animals sink to the bottom, feeding the fish, crabs, brittle stars, mollusks, and other invertebrates that inhabit the seafloor.

As permanent ice diminishes, a new ocean is opening up. Scientists are just beginning to gather data about the biology of the Central Arctic Ocean because, until now, it has been difficult to access. Some of the preliminary results presented in the booklet illustrate that this emerging ocean is teeming with life and connected to the rest of the world's oceans.

Pew's international Arctic campaign is working with Arctic countries, scientists, the fishing industry, and indigenous peoples to support an agreement that would protect the international waters of the Central Arctic Ocean and its living marine resources from unregulated or unsustainable commercial fishing.

Courtesy of The Pew Charitable Trusts, 2013.

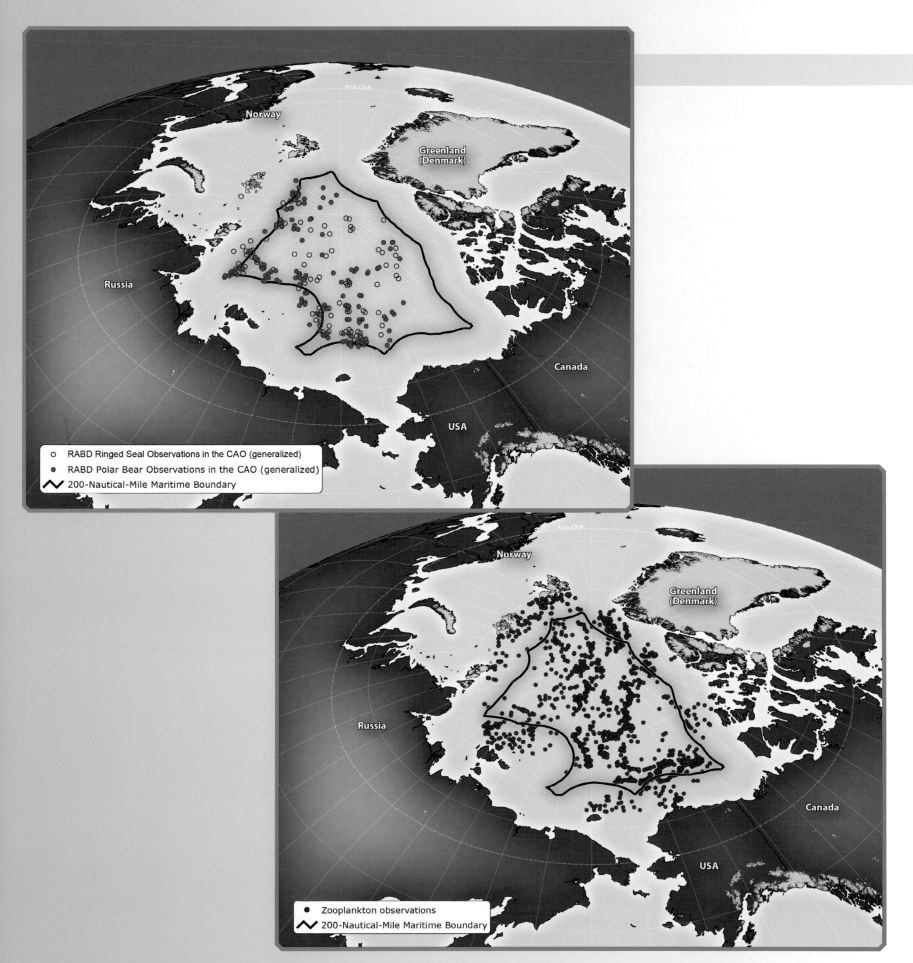

RABD Ringed Seal Observations in the CAO (generalized)
RABD Polar Bear Observations in the CAO (generalized)
200-Nautical-Mile Maritime Boundary

Zooplankton observations
200-Nautical-Mile Maritime Boundary

Central Gulf of Mexico Facilities and Lease Map

Enterprise Products Partners LP

Houston, Texas, USA

By Stephen J. Visconti

Contact

Stephen J. Visconti, SVisconti@eprod.com

Software

ArcGIS 10.1 for Desktop

Data Sources

Texas Railroad Commission, Bureau of Ocean Energy Management, MapSearch, Energy Graphics, Enterprise databases

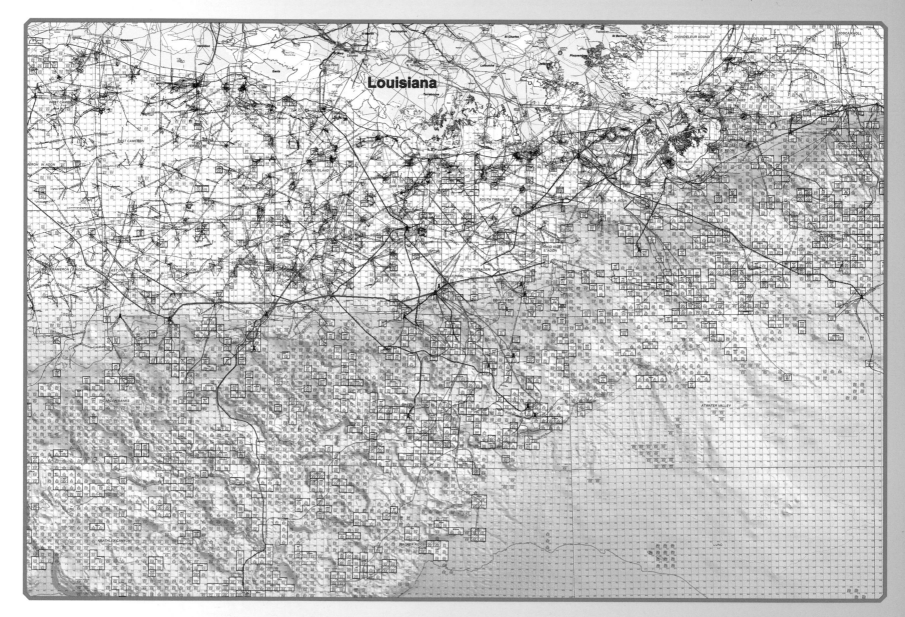

This map is used in business and commercial project creation and evaluation and can be a starting point to evaluate offshore leasing and drilling activity. The activity of operators of interest can be highlighted in various concession areas where existing corporate-owned pipeline infrastructure may enhance the ability to move new reserve discoveries to company-owned processing plants onshore. Offshore leasing and drilling activity has been upgraded in-house (project outlines and names, leases with plans of exploration or discoveries, etc.) using various industry data publications and Internet-posted operator presentations.

Courtesy of 2012 Enterprise Products Partners LP

Field Development 2012 Survey of Arctic and Cold Region Technology for Offshore

INTECSEA

Houston, Texas, USA

By John R. Brand

Contact

John R. Brand, john.brand@intecsea.com

Software

ArcGIS 10.0 for Desktop

Data Sources

US National Snow and Ice Data Center, National Oceanic and Atmospheric Administration National Geophysical Data Center, US Naval Ocean Office, Arctic Studies Program, Natural Resources Canada

Innovative technologies are required to develop oil and gas projects in the Arctic and other cold regions. INTECSEA, with experience in the Canadian, United States, and Russian Arctic, has advanced hostile, cold-climate engineering technology for the safe development of both existing and future oil and gas fields in the Arctic. The company developed this map to assist Arctic technology innovation by establishing a geospatial database of engineering technologies used at existing Arctic project locations. The map uses ArcGIS to gather, collate, and associate Arctic geographic, environmental, and sea ice data to the engineering technologies required to safely develop and transport produced hydrocarbons from these existing Arctic projects to market. The map also supports Arctic engineering specialists in performing pipeline, production, and hydrocarbon transportation technology assessments in Arctic lease areas to identify what new technologies will be required to safely develop future projects in the Arctic region.

Courtesy of INTECSEA.

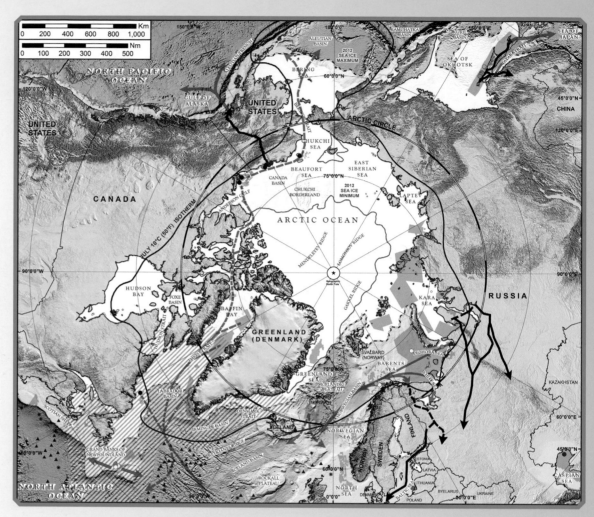

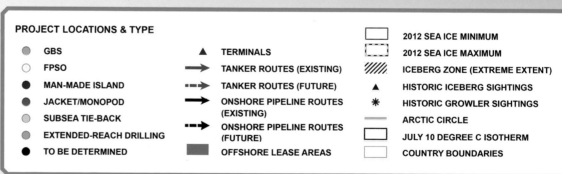

PROJECT LOCATIONS & TYPE

- 🔵 GBS
- ⚪ FPSO
- ⚫ MAN-MADE ISLAND
- 🔴 JACKET/MONOPOD
- 🟢 SUBSEA TIE-BACK
- 🔵 EXTENDED-REACH DRILLING
- ⚫ TO BE DETERMINED

- ▲ TERMINALS
- → TANKER ROUTES (EXISTING)
- ⇢ TANKER ROUTES (FUTURE)
- ➡ ONSHORE PIPELINE ROUTES (EXISTING)
- ⇢ ONSHORE PIPELINE ROUTES (FUTURE)
- ▬ OFFSHORE LEASE AREAS

- ☐ 2012 SEA ICE MINIMUM
- ☐ 2012 SEA ICE MAXIMUM
- ▨ ICEBERG ZONE (EXTREME EXTENT)
- ▲ HISTORIC ICEBERG SIGHTINGS
- ✳ HISTORIC GROWLER SIGHTINGS
- ⎯ ARCTIC CIRCLE
- ☐ JULY 10 DEGREE C ISOTHERM
- ☐ COUNTRY BOUNDARIES

Pipeline System 1 Schematic

Denbury Resources Inc.

Plano, Texas, USA

By Brian Reeder and Phil Roybal

Contact

Phil Roybal, philip.roybal@denbury.com

Software

ArcGIS 10.1 for Desktop, ArcGIS Schematics

Data Source

Denbury Resources Inc.

This is an example of a schematic representation of an interstate transmission pipeline system, production sites, and compressor station sites. The schematic map was created using the ArcGIS Schematics extension from pipeline GIS data. The standard schematic template Geo-Initial Positions was used, and then some cleanup and labeling was used to get the finished layout for the map.

Courtesy of Denbury Resources Inc.

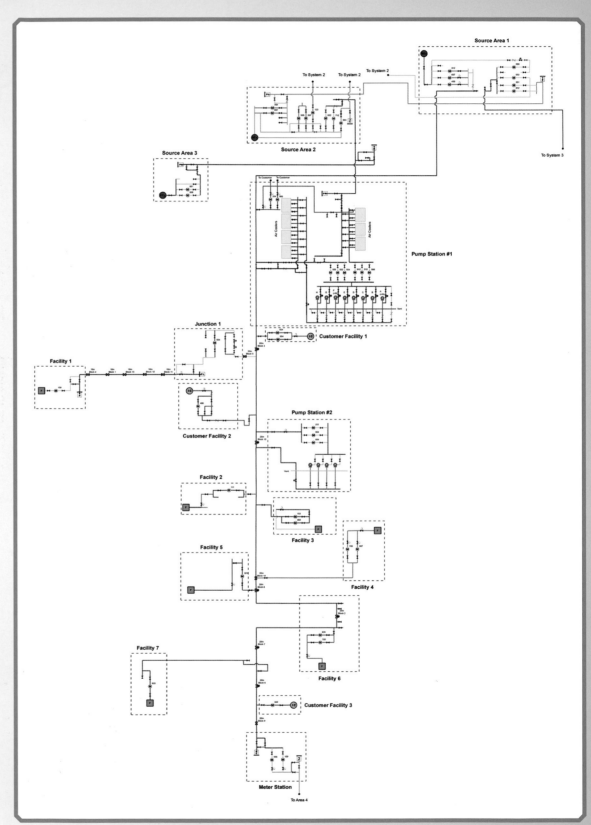

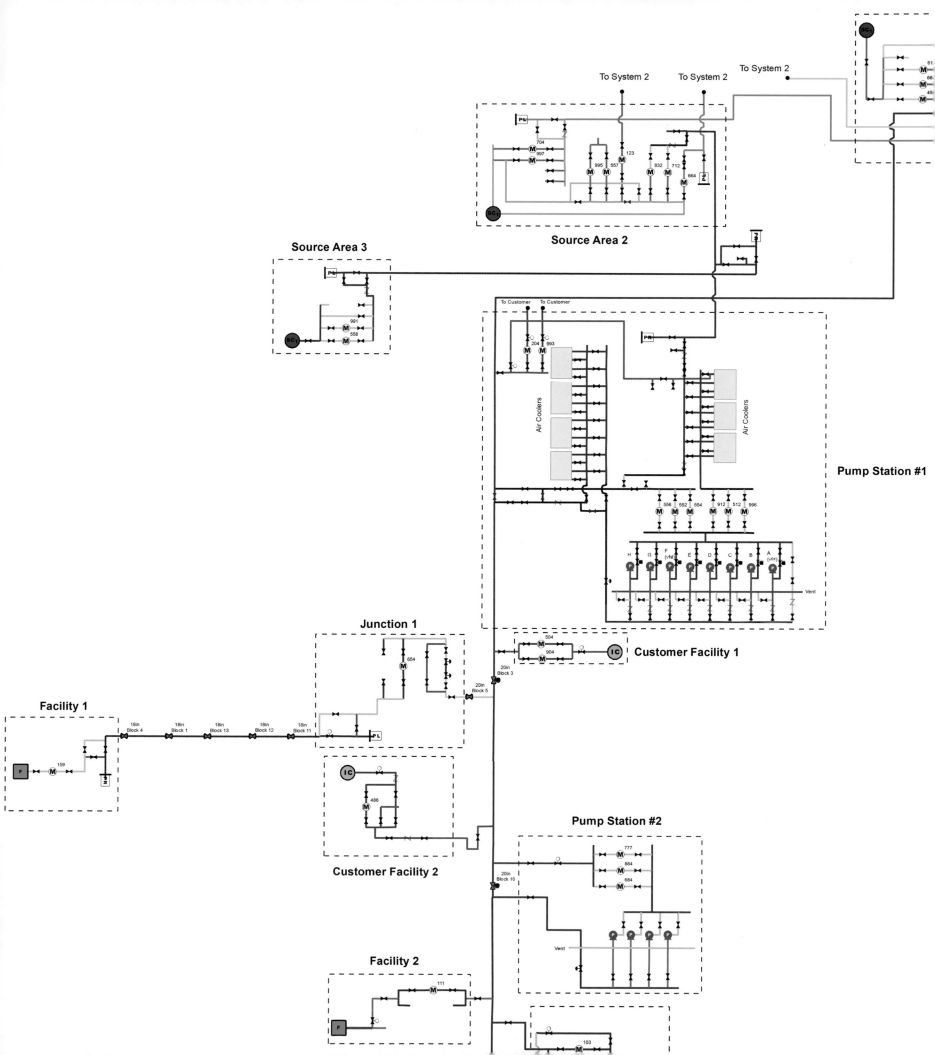

An Evaluation of Six Groundwater Quality Parameters

New Jersey Department of Environmental Protection

Trenton, New Jersey, USA

By Nick Procopio, Judy Louis, and Tom Atherholt

Contact

Nick Procopio, nick.procopio@dep.state.nj.us

Software

ArcGIS 10.0 for Desktop

Data Sources

New Jersey Department of Environmental Protection, New Jersey Private Well Testing Act

This map was developed to provide a visual assessment of the variability of certain water-quality parameters in New Jersey's groundwater. New Jersey's Private Well Testing Act (PWTA) was passed in March 2001, and sampling commenced in September 2002. It is the nation's first statewide program and among only a few state-run programs. In an effort to protect human health, the PWTA requires that source water from private wells be tested before the closing of any real estate transaction or new lease agreement. The act requires the testing of various water-quality parameters, thirty-two of which are primary parameters of human health concern along with three non-health-related secondary parameters. The data is electronically submitted to the New Jersey Department of Environmental Protection (NJDEP) by the buyer- or seller-contracted laboratories.

The NJDEP uses the data collected to evaluate groundwater quality throughout the state and to inform municipalities, counties, and other government entities of potential hazards. The number of individual wells sampled throughout the state varies by parameter. These maps present the data of three primary (nitrate/nitrite, arsenic, and radionuclides) and three secondary (pH, iron, and manganese) parameters, all naturally occurring, from samples collected between September 2002 and January 2011. The results show the variability in the concentration of each parameter throughout New Jersey or specific region where its analysis is required.

Courtesy of Nick Procopio, New Jersey Department of Environmental Protection 2013.

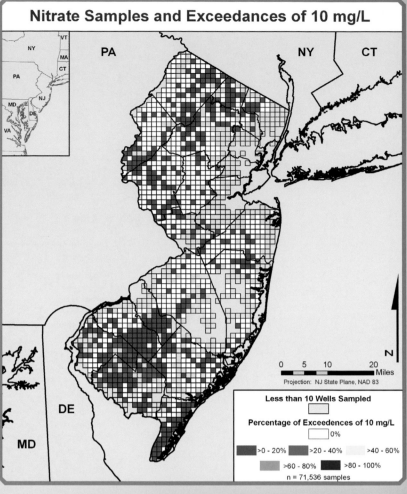

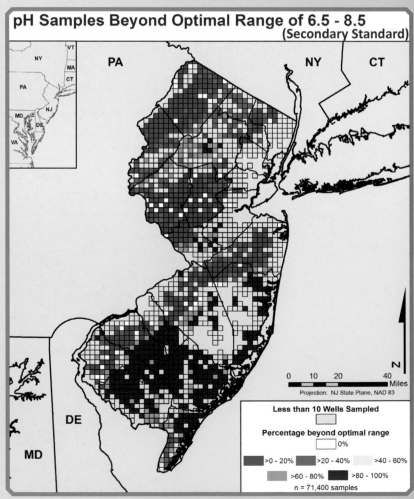

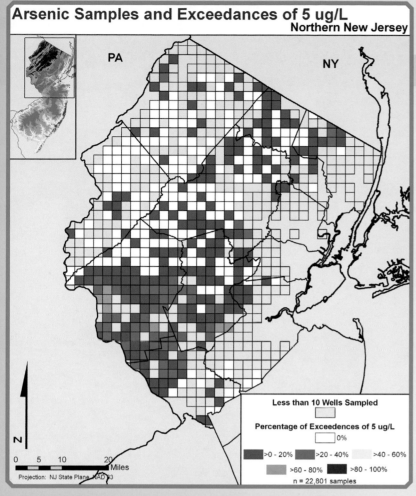

Arsenic Samples and Exceedances of 5 ug/L
Northern New Jersey

PA
NY

N

0 5 10 20 Miles

Projection: NJ State Plane, NAD 83

Less than 10 Wells Sampled

Percentage of Exceedances of 5 ug/L
- 0%
- >0 - 20%
- >20 - 40%
- >40 - 60%
- >60 - 80%
- >80 - 100%

n = 22,801 samples

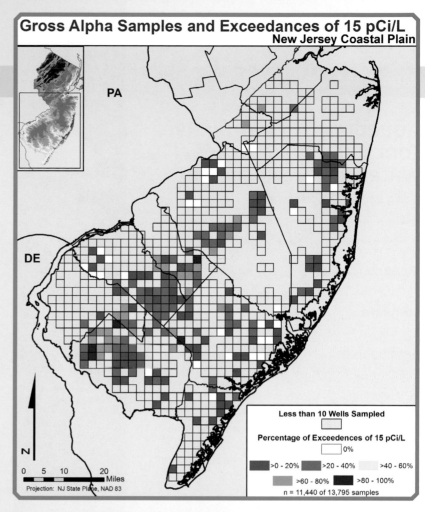

Gross Alpha Samples and Exceedances of 15 pCi/L
New Jersey Coastal Plain

PA
NY
DE

N

0 5 10 20 Miles

Projection: NJ State Plane, NAD 83

Less than 10 Wells Sampled

Percentage of Exceedances of 15 pCi/L
- 0%
- >0 - 20%
- >20 - 40%
- >40 - 60%
- >60 - 80%
- >80 - 100%

n = 11,440 of 13,795 samples

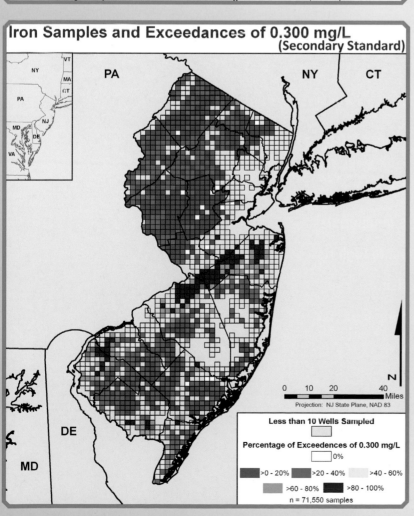

Iron Samples and Exceedances of 0.300 mg/L
(Secondary Standard)

NY VT
PA MA
CT
NY
PA
MD NJ
DE
VA

0 10 20 40 Miles

N

Projection: NJ State Plane, NAD 83

MD
DE

Less than 10 Wells Sampled

Percentage of Exceedances of 0.300 mg/L
- 0%
- >0 - 20%
- >20 - 40%
- >40 - 60%
- >60 - 80%
- >80 - 100%

n = 71,550 samples

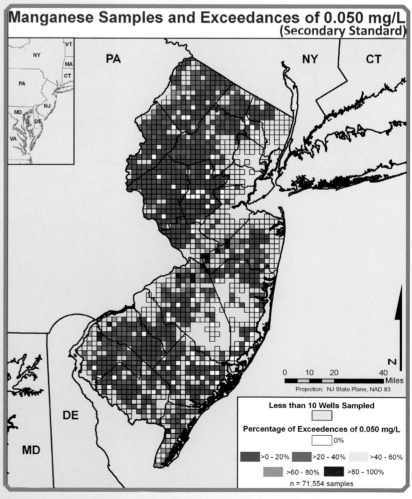

Manganese Samples and Exceedances of 0.050 mg/L
(Secondary Standard)

NY VT
PA MA
CT
NY
PA
MD NJ
DE
VA

0 10 20 40 Miles

N

Projection: NJ State Plane, NAD 83

MD
DE

Less than 10 Wells Sampled

Percentage of Exceedances of 0.050 mg/L
- 0%
- >0 - 20%
- >20 - 40%
- >40 - 60%
- >60 - 80%
- >80 - 100%

n = 71,554 samples

Aqueduct: An Interactive Tool to Empower Global Water Risk Assessment

World Resources Institute

Washington, DC, USA

By Francis Gassert, Paul Reig, and Nick Price

Contact

Francis Gassert, fgassert@wri.org

Software

ArcGIS 10.1 for Desktop, Adobe Illustrator, Adobe InDesign

Data Source

World Resources Institute

The World Resources Institute's Aqueduct open online mapping tool creates customized water risk maps of the world and lets users compare risk exposure across the locations that matter most to them. This map combines data from Aqueduct's twelve different indicators to create a comprehensive look at where water risks are greatest around the world. Each of Aqueduct's indicators is mapped, and each map tells a piece of the complex story of water risk, from water stress to drought severity to access to drinking water. By combining data on all these indicators, Aqueduct simplifies complex hydrological data into a clear and compelling map.

Courtesy of World Resources Institute.

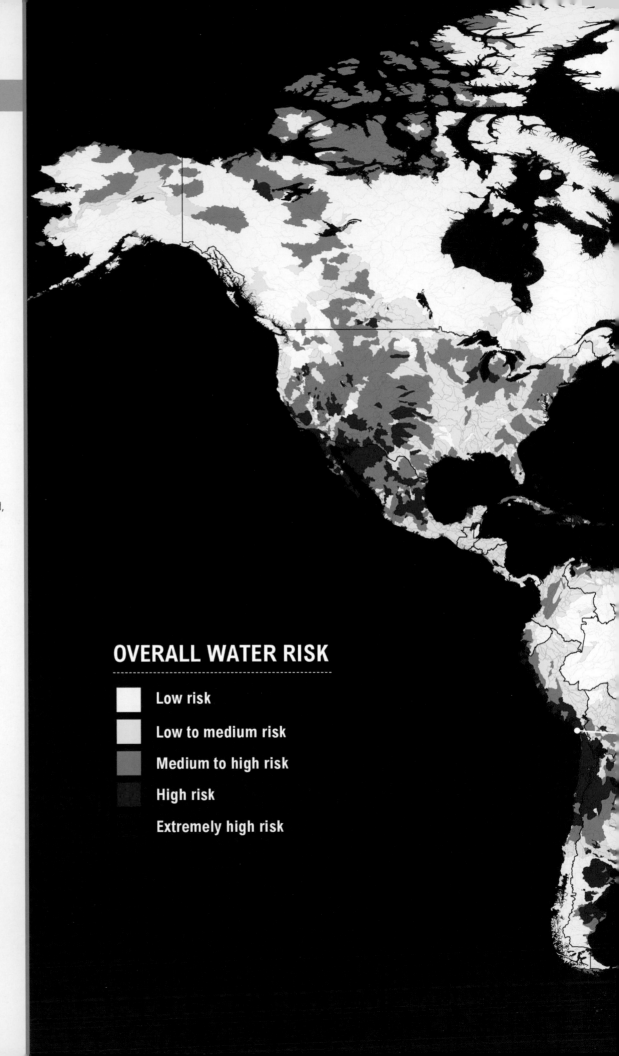

OVERALL WATER RISK

Low risk

Low to medium risk

Medium to high risk

High risk

Extremely high risk

BASELINE WATER STRESS

THREATENED AMPHIBIANS

GROUNDWATER STRESS

FLOOD OCCURRENCE

US Drought Risk Map: A Cumulative Approach from 2000 to 2012

***Texas Tech University and Universidade Federal de Campina Grande
(Federal University of Campina Grande, Brazil)***
Lubbock, Texas, USA

By Iana Rufino, Kevin Mulligan, Lucia Barbato, and Santosh Seshadri

Contact

Iana Rufino, iana.alexandra@ufcg.edu.br

Software

ArcGIS 10.1 for Desktop

Data Sources

National Drought Mitigation Center, University of Nebraska-Lincoln, US Department of Agriculture, National Oceanic and Atmospheric Administration

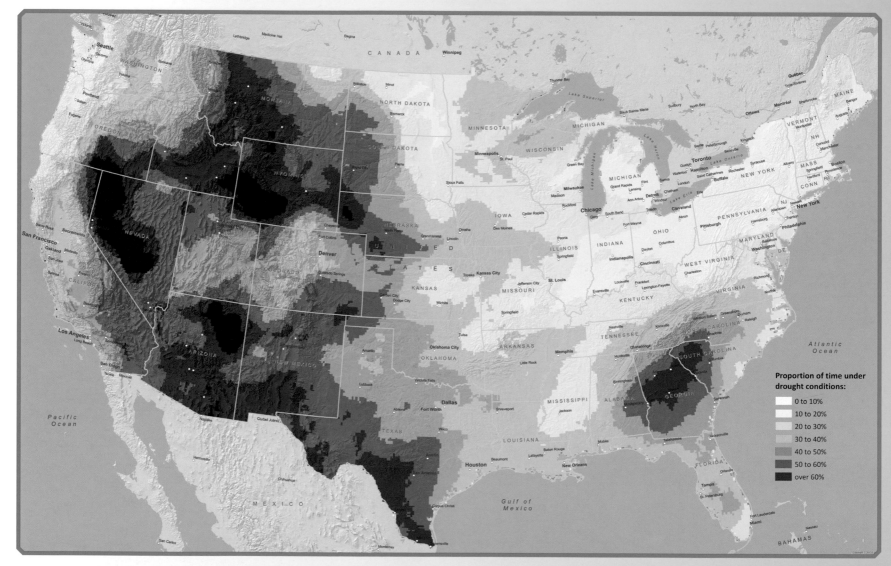

Proportion of time under drought conditions:
- 0 to 10%
- 10 to 20%
- 20 to 30%
- 30 to 40%
- 40 to 50%
- 50 to 60%
- over 60%

This map shows drought risk expressed as the percentage of time an area was under drought from 2000 through 2012. To produce the map, drought conditions were defined using the Palmer Hydrological Drought Index (PHDI), a long-term drought index developed to quantify the impact of drought on hydrological conditions such as reservoir and groundwater levels. Compared to most other drought indexes, the PHDI responds much more slowly to changes in weather.

To create the drought risk map, 678 weekly PHDI maps were downloaded from the US Drought Monitor in shapefile format. The polygon features in each shapefile delineate areas that were under drought and the severity of the drought. ModelBuilder was used to process and combine the 678 shapefiles into a single cumulative drought raster. This drought raster was classified based on the proportion of time an area was under drought conditions. For example, in the model, a grid cell value of 678 would represent 100 percent of the thirteen-year period under drought.

Using this approach, the map shows those areas of the United States that experienced cumulative hydrological drought from 2000 through 2012. While it might be expected that many of the arid and semiarid regions of the western United States are at risk to frequent and prolonged drought, parts of the southeastern United States experienced drought conditions for a high percentage of the time.

Courtesy of Center for Geospatial Technology.

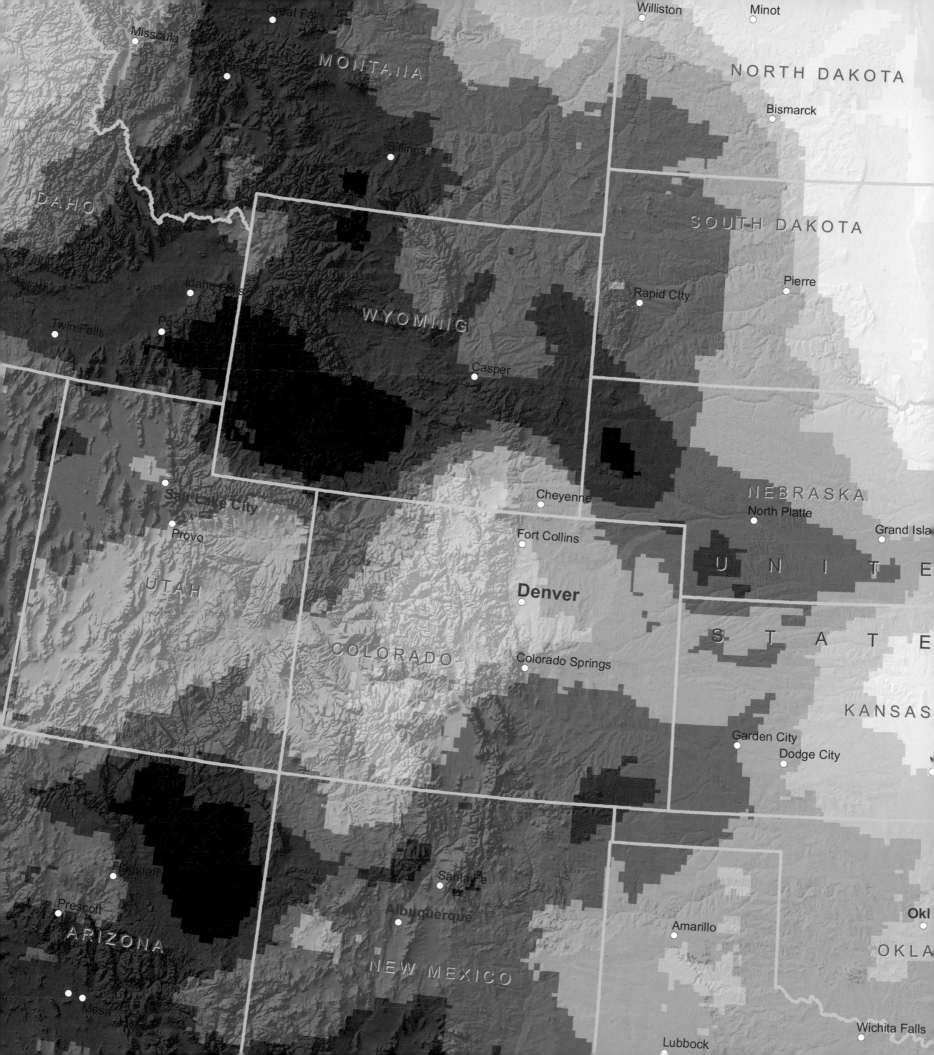

Zinc Runoff Concentration in Los Angeles

Cal Poly University, Pomona

Pomona, California, USA

By Pedro Ventura, Phillip Park, Mary Cadena, and John Alvarez

Contact

Lin Wu, lwu@csupomona.edu

Software

ArcGIS 10.0 for Desktop, Adobe Photoshop, Adobe Illustrator

Data Sources

Los Angeles County GIS Data Portal, City of Los Angeles

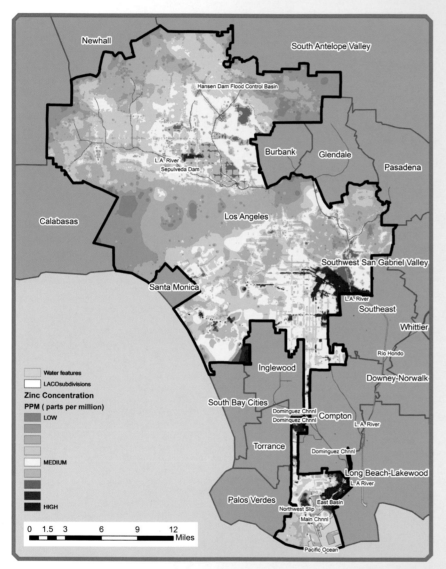

Zinc is one of the many pollutants found in runoff that has been proven to be problematic to waterways in urban areas such as Los Angeles. It is found in higher traceable concentrations compared to other metal pollutants in runoff produced during rainfall events and can be an indicator of greater pollution in an area. This map of Los Angeles demonstrates an effort to determine where higher concentrations of zinc may occur. Taking into account information pertaining to parcel land use, soil permeability, rainfall amount, and runoff produced, zinc hot spots within Los Angeles emerged.

It was apparent that areas with heavy industrial land use along the Los Angeles River generated higher pollutants. The study results also identified areas of higher zinc concentrations toward the San Fernando Valley dominated by residential and commercial land uses. This indicates the possibility that nonindustrial areas are capable of yielding high amounts of metal contaminants. The pattern further reinforces the hot spot analysis pattern examined for the concentrations of zinc. The results provided a way to visually understand the behavior of metal pollutants and the possible effects to Los Angeles waterways such as the Los Angeles River.

Courtesy of Cal Poly University, Pomona.

Energy Planning for Noninterconnected Zones in Colombia Based on Multicriteria Analysis Using GIS

Universidad de los Andes (University of the Andes)

Bogota DC, Cundinamarca, Colombia

By Lised Chaves, Jonathan Villota, Daniel Paez, and Ángela Cadena

Contact

Lised Chaves, lk.chaves1128@uniandes.edu.co

Software

ArcGIS 10.1 for Desktop

Data Sources

Mining and Energy Planning Unit, GIS for planning and territorial management, Universidad de los Andes

The lack of access to energy affects people's quality of life, including health, education, productivity, and economic growth. Currently in Colombia, there are approximately 1.9 million people without power service, representing the largest population with an unsatisfied basic needs (UBN) index and low-income levels.

These maps are part of a GIS energy planning model to select priority areas for electric generation facilities. The analysis is based on multicriteria involving demand; supply energy (solar photovoltaic, wind); infrastructure; proximity to diesel plants; and social, economic, and environmental factors. The analysis is performed with a priority on Colombian territories that are not connected to the National Electricity Transmission System and where a large availability of physical resources makes alternative technologies more competitive.

Courtesy of Universidad de los Andes.

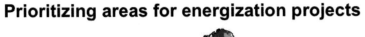

Prioritizing areas for energization projects

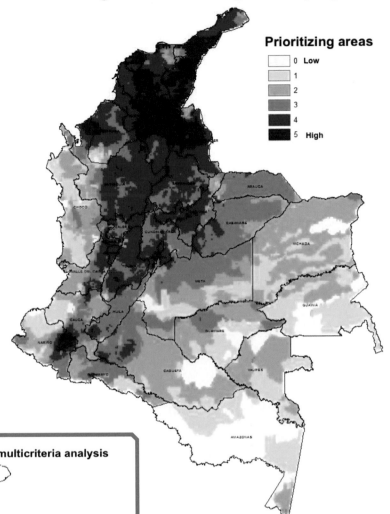

Prioritizing areas

- 0 Low
- 1
- 2
- 3
- 4
- 5 High

Localities with greater weight in multicriteria analysis

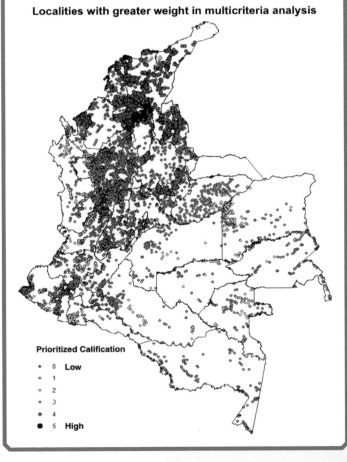

Prioritized Calification

- 0 Low
- 1
- 2
- 3
- 4
- 5 High

Senior Shedding

Portland State University
Portland, Oregon, USA

By Richard Lycan and Charles Rynerson

Contact
Richard Lycan, lycand@pdx.edu

Software
ArcGIS 10.0 for Desktop, Adobe Illustrator CS6, Microsoft Excel

Data Sources
Oregon Health Division, US Census, Portland Metro's Regional Land Information System, US Geological Survey digital elevation models

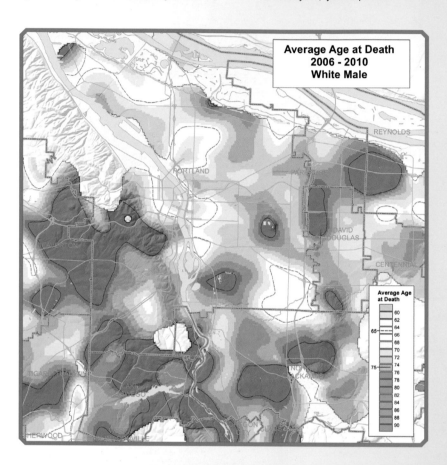

Average Age at Death 2006 - 2010 White Male

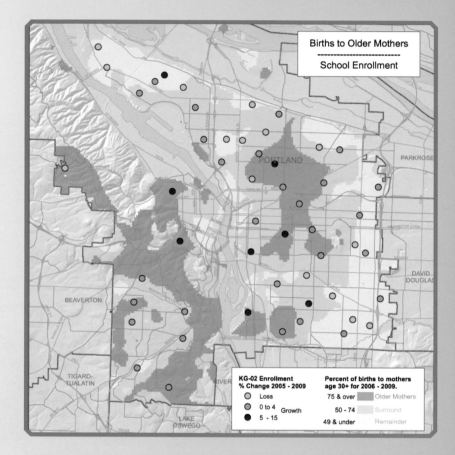

Births to Older Mothers — School Enrollment

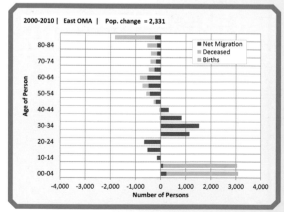

Portland State University's Population Research Center (PRC) regularly conducts demographic studies for school districts. The most commonly requested projects are for enrollment forecasts for school districts, or for a district and its individual schools. This map was part of the PRC's enrollment forecasting work for Portland Public Schools to understand why school enrollment was growing in neighborhoods with older birth mothers. Mapping of birth record data showed that the highest proportion of births to mothers aged 30 and over was concentrated in a few areas of the district with older homes previously inhabited by seniors. Most of the vacancies were provided by senior households through mortality or out-migration or "senior shedding."

The research reflected in these maps was an outgrowth of the Population Research Center's enrollment forecasting work for the Portland School District. School enrollment was growing in neighborhoods with older birth mothers and the research focused on the source of the housing vacancies for them. The highest proportion of births to mothers age 30 and over was concentrated in a few areas of the district, shown in blue on the map. Many of these families with older birth mothers were drawn by the vintage Craftsman-style houses previously inhabited by seniors.

The graph shows that for the east side older mothers area, the largest number of in-migrants were from their late 20s to mid- 30s, along with their children. On the maps, deaths per acre mainly reflect the distribution of the senior population. Migrants per acre have a more complex pattern with a few areas where senior housing is concentrated showing large increases but broad areas showing losses through out-migration, or senior shedding. A map created from geocoded death record data shows how longevity varies locally for white males from under 65 years to over 80 years.

Courtesy of Portland State University.

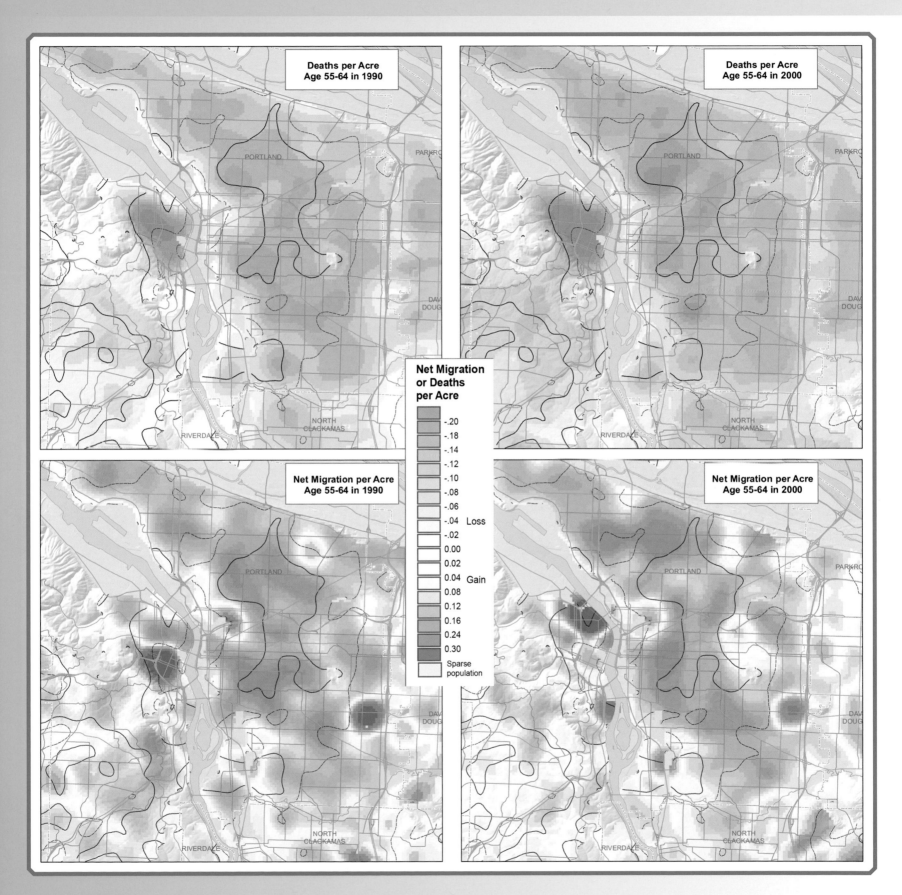

Deaths per Acre
Age 55-64 in 1990

Deaths per Acre
Age 55-64 in 2000

Net Migration per Acre
Age 55-64 in 1990

Net Migration per Acre
Age 55-64 in 2000

Net Migration
or Deaths
per Acre

-.20
-.18
-.14
-.12
-.10
-.08
-.06
-.04   Loss
-.02
0.00
0.02
0.04   Gain
0.08
0.12
0.16
0.24
0.30
Sparse
population

Hurricane Sandy Waterway Debris Removal Project

New Jersey Department of Environmental Protection

Trenton, New Jersey, USA

By Edward Apalinski

Contact

Edward Apalinski, edward.apalinski@dep.state.nj.us

Software

ArcGIS 10.0 for Desktop, ArcGIS Online

Data Sources

Federal Emergency Management Agency, New Jersey Department of Environmental Protection, US Fish and Wildlife Service

The efforts of the New Jersey Department of Environmental Protection for the Hurricane Sandy Waterway Debris Removal Project began with organizing the impacted waterways into eleven debris management zones. These waterways included coastal, tidal, and wetlands from Bergen County to Cape May County and up the Delaware Bay to the Delaware Memorial Bridge in Salem County.

The next step identified, acquired, and mapped twenty-plus GIS data layers for state and federal environmental laws and executive orders, which must be accounted for during operations for Federal Emergency Management Agency (FEMA) reimbursement. These include sensitive wildlife areas, aquaculture, natural and historic resources, wild and scenic rivers, federal and state channels, and wetlands.

The last step was to share the map with all interested parties. In March 2013, a team of contractors began removing waterway debris including cars, vessels, docks, and boardwalks. To maintain FEMA eligibility, the sensitive areas from the previous step must be accounted for during field operations. Production of an ArcGIS Online map allowed the datasets in the web map to be viewed in the office for mission planning and in the field via mobile device to identify sensitive areas. When the project ended on October 30, 2013, more than 101,000 cubic yards of debris had been removed from nearly 195,000 acres of waterways.

Courtesy of New Jersey Department of Environmental Protection, Bureau of GIS.

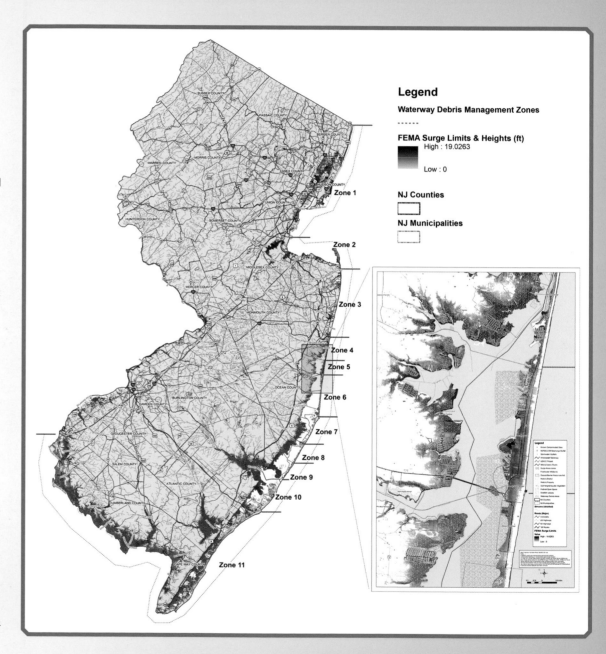

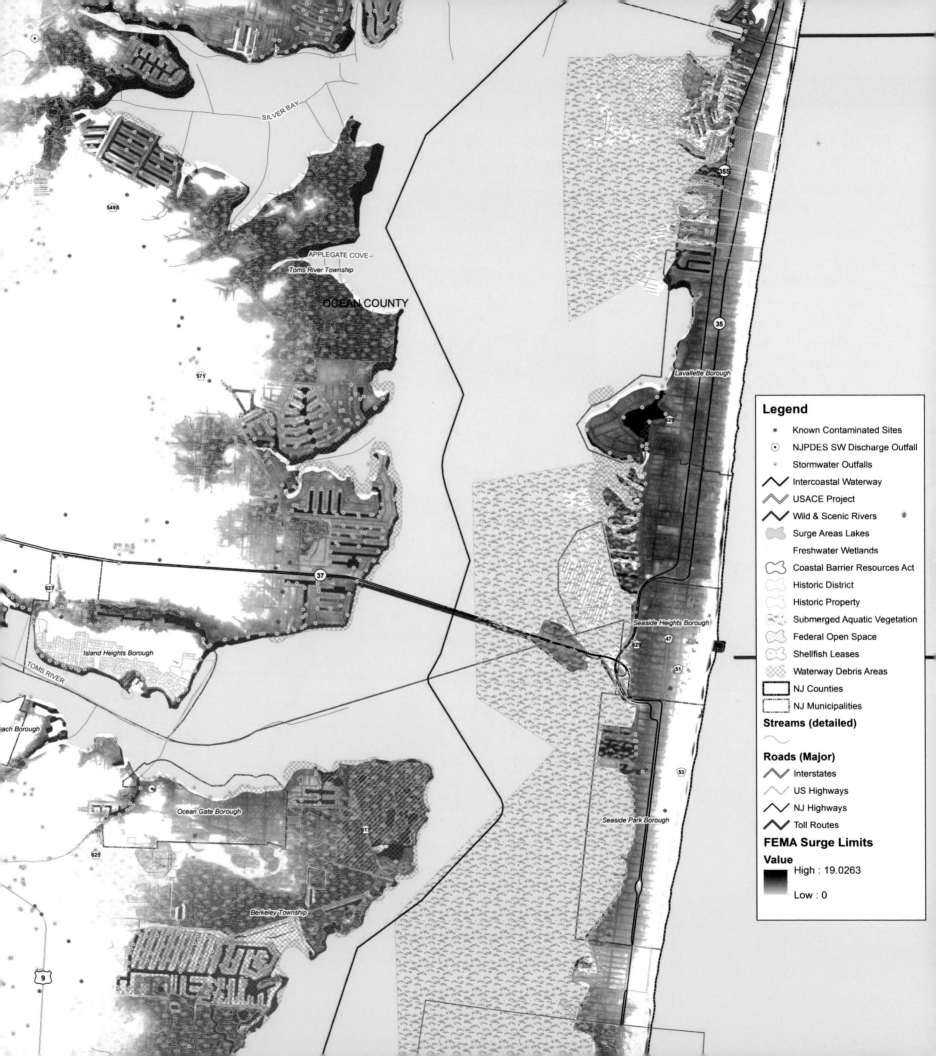

Legend

- • Known Contaminated Sites
- ⊙ NJPDES SW Discharge Outfall
- • Stormwater Outfalls
- ⌒ Intercoastal Waterway
- ⌒ USACE Project
- ⌒ Wild & Scenic Rivers
- Surge Areas Lakes
- Freshwater Wetlands
- Coastal Barrier Resources Act
- Historic District
- Historic Property
- Submerged Aquatic Vegetation
- Federal Open Space
- Shellfish Leases
- Waterway Debris Areas
- ☐ NJ Counties
- ☐ NJ Municipalities

Streams (detailed)

Roads (Major)
- ⌒ Interstates
- ⌒ US Highways
- ⌒ NJ Highways
- ⌒ Toll Routes

FEMA Surge Limits

Value
High : 19.0263

Low : 0

SILVER BAY

APPLEGATE COVE

Toms River Township

OCEAN COUNTY

Lavallette Borough

Island Heights Borough

TOMS RIVER

...ach Borough

Ocean Gate Borough

Seaside Heights Borough

Seaside Park Borough

Berkeley Township

549S

571

627

37

625

917

9

35S

35

629

629

47

51

53

Land Acquisition/Regeneration and Implementation—Types of Affected Plots, Unplanned Settlements

AECOM Middle East

Abu Dhabi, United Arab Emirates

By Sumanth Toranala

Contact

Sumanth Toranala, Sumanth.tornala@aecom.com

Software

ArcGIS for Desktop

Data Sources

AECOM—London, Arabia

AECOM carries out a range of planning, design, and engineering projects aimed at transforming selected unplanned settlements into revitalized, safe, and complete communities and neighborhoods and facilitating and responding to regional and local storm water management interventions.

In 2009 and 2011, the southern and western parts of Jeddah, Saudi Arabia, experienced disastrous floods, resulting in the loss of life and significant infrastructure and property damage.

The intervention program aimed to minimize the probability of fatalities and reduce the amount of property damage and traffic and nuisance disruptions created during a flooding event. The two-and-a-half-year program commenced in June 2011.

The maps show numerous ad hoc areas identified throughout the city as essentially "unplanned settlements." These neighborhoods experience many social problems, exhibit a poorly built form and land-use balance, and do not have any form of organized infrastructure or storm water management systems. The haphazard nature of these localities exacerbated the impacts of flooding.

The goal is to improve safety and security, manage storm water, and deliver public betterment. The project aims to transform the unplanned settlements into safe, revitalized, and complete communities and to avoid unnecessary intervention, property expropriation, and disruption.

Courtesy of Sumanth Toranala, AECOM Middle East.

Commercial
General Services
Government/Other
Industrial
Residential - Commercial
Residential
Store
Transportation
Vacant

Existing Condition

Commercial - Retail
Education
Government
PetrolStation - Existing
Open Space
Baraha
Community Facilities
Power Network
Sub Station
Residential
Sikkah
Tanks _ Reservoir
Right Of Way

Proposed AECOM Land Use

- Area of plot affected by ROW (acquisition required)
- Area of plot to be acquired for other purposes
- Area of plot partially affected by ROW (acquisition not required, greater than 100 sqm inside residential block)

Existing Plots and Proposed Land Use

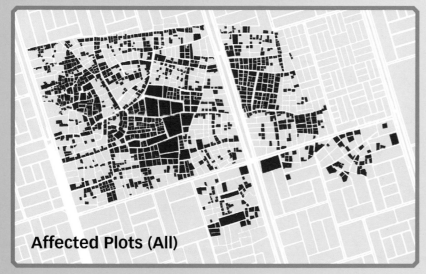

Affected Plots (All)

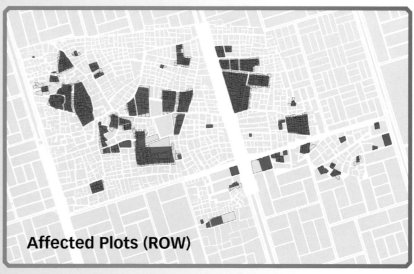

Affected Plots (ROW)

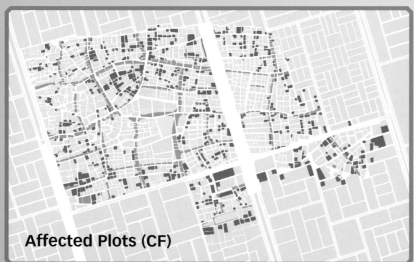

Affected Plots (CF)

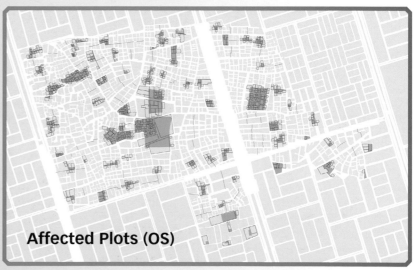

Affected Plots (OS)

Optimizing Measurement of 4G LTE Broadband Access in Virginia

Virginia Tech
Blacksburg, Virginia, USA

By Erica E. Adams

Contact

Erica E. Adams, vterica@gmail.com

Software

ArcGIS 10.1 for Desktop

Data Sources

US Census Bureau; Oak Ridge National Laboratory; National Broadband Map (NTIA); Virginia Geographic Information Network

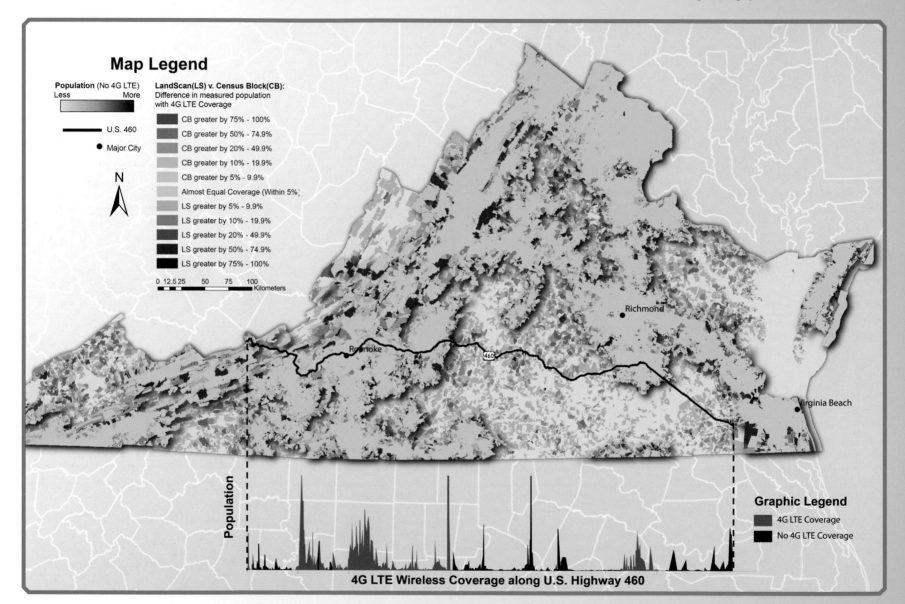

Map Legend

Population (No 4G LTE)
Less — More

U.S. 460

● Major City

N

LandScan(LS) v. Census Block(CB):
Difference in measured population with 4G LTE Coverage

- CB greater by 75% - 100%
- CB greater by 50% - 74.9%
- CB greater by 20% - 49.9%
- CB greater by 10% - 19.9%
- CB greater by 5% - 9.9%
- Almost Equal Coverage (Within 5%)
- LS greater by 5% - 9.9%
- LS greater by 10% - 19.9%
- LS greater by 20% - 49.9%
- LS greater by 50% - 74.9%
- LS greater by 75% - 100%

0 12.5 25 50 75 100 Kilometers

Richmond

Roanoke

460

Virginia Beach

Population

Graphic Legend
- 4G LTE Coverage
- No 4G LTE Coverage

4G LTE Wireless Coverage along U.S. Highway 460

Oak Ridge National Laboratory created the LandScan ambient population dataset to represent average population distribution over twenty-four hours. The model combines multiple population-relevant datasets and weights them to produce a one-kilometer resolution population distribution raster. This analysis compares the use of US Census population data at the block level and LandScan's ambient population model to calculate the percentage of the population with 4G LTE broadband access.

This analysis shows that at the census block level, these datasets are comparable for use in analysis of broadband coverage: only 4 percent of census blocks covered by 4G LTE differed by more than 5 percent when measuring population using LandScan and census data, representing less than 2 percent of the covered population (Census: 88.7 percent; LandScan: 86.9 percent).

Broadband data for the analysis was collected through the Virginia State Broadband Initiative, which was funded through the National Telecommunications and Information Administration in an effort to realize the purposes of the American Recovery and Reinvestment Act and the Broadband Data Improvement Act. The Center for Geospatial Information Technology leverages innovative geospatial solutions for broadband mapping and analysis and many other research areas.

Courtesy of Erica Adams, Center for Geospatial Information Technology, Virginia Tech.

The Grey-Bruce Map

County of Bruce

Wiarton, Ontario, Canada

By Paulette Weber, Justin Kraemer, and Grant Nicholson

Contact

Justin Kraemer, jkraemer@brucecounty.on.ca

Software

ArcGIS 10.0 for Desktop, Adobe Illustrator CS5.1, Adobe InDesign CS5.5

Data Sources

County of Bruce, County of Grey, Bruce Trail Conservancy, Ontario Ministry of Natural Resources

Tourism is the number-two industry in both Grey and Bruce Counties, with over 3.2 million visitors drawn each year to experience the beaches, trails, lighthouses, waterfalls, national parks, and stunning views. The annual economic receipts exceed $390 million.

With an area of nearly 8,600 square kilometers (3,400 square miles), a quality map is essential for people to find their way. Until the late 1990s, a 1:400,000-scale visitor map was published every year by the independent Grey-Bruce Tourism Association. While very popular and well used, it had shortcomings and soon disappeared.

By 2006, a replacement was sought. From evaluation of the old Grey-Bruce Map, county tourism and GIS staff gained many ideas for how to not only bring it back but to make it far better. The map came together with spatial data maintained by both counties and other members of the Ontario Geospatial Data Exchange. The volume of content demanded a larger scale of 1:190,000, allowing for rich detail like the labeling of almost every road, water body, and point of interest.

The most distinguishing improvement of this generation of the Grey-Bruce Map is the shaded relief elevation that adds not only to visual appeal but offers an accurate impression of regional topography. The Beaver Valley clearly dominates, cutting halfway into Grey County. The Niagara Escarpment conspicuously snakes across both counties. The highest elevations in the southeast are visibly apparent.

This project marked the first time such an abundance of map detail had been compiled on a regional scale.

Re-creating the 1902 Railroad Map of Vermont

Vermont Agency of Transportation

Montpelier, Vermont, USA

By Stephen Smith

Contact

Stephen Smith, stephen.smith@state.vt.us

Software

ArcGIS 10.1 for Desktop

Data Sources

Vermont Agency of Transportation, map data © OpenStreetMap contributors, CC-BY-SA, US Geological Survey National Hydrography Dataset, Massachusetts Department of Transportation, Esri/TANA

Due to a labor dispute, the Rutland Railroad Company went bankrupt in 1961. To prevent the railroad from becoming abandoned, the State of Vermont purchased approximately 125 miles of rights-of-way within Vermont and New York. The Vermont Agency of Transportation (VTrans) now manages the assets and the property, while operation is leased to Vermont Rail Systems.

The railroad has played a major part in Vermont's history. In honor of this tradition, the VTrans Rail Section created this map in the style of a 1902 Railroad Commissioners Report Map. Working from a high-resolution scan provided by Middlebury College's digital map collection, this map mimics the style of the old map down to the faded color of the page.

Using current GIS data on active rail lines and rail trails, this map reflects 113 years of changes to Vermont's rail infrastructure and the enormous evolution of cartographic tools over that time period. It is challenging to communicate the soft, human touch of a pen stroke with crisp, perfect vector lines. Remembering the styles of old helps recapture some of this lost grace.

Courtesy of Vermont Agency of Transportation.

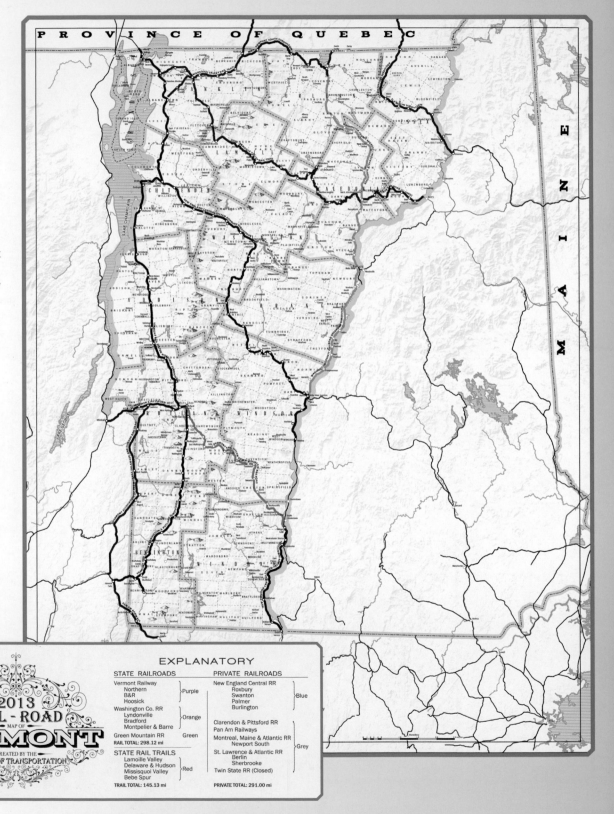

EXPLANATORY

STATE RAILROADS

Vermont Railway
  Northern
  B&R — Purple
  Hoosick
Washington Co. RR
  Lyndonville
  Bradford — Orange
  Montpelier & Barre
Green Mountain RR — Green

RAIL TOTAL: 298.12 mi

STATE RAIL TRAILS

  Lamoille Valley
  Delaware & Hudson
  Missisquoi Valley — Red
  Bebe Spur

TRAIL TOTAL: 145.13 mi

PRIVATE RAILROADS

New England Central RR
  Roxbury
  Swanton — Blue
  Palmer
  Burlington

Clarendon & Pittsford RR
Pan Am Railways
Montreal, Maine & Atlantic RR
  Newport South — Grey
St. Lawrence & Atlantic RR
  Berlin
  Sherbrooke
Twin State RR (Closed)

PRIVATE TOTAL: 291.00 mi

2013 RAIL-ROAD MAP OF VERMONT — CREATED BY THE AGENCY OF TRANSPORTATION

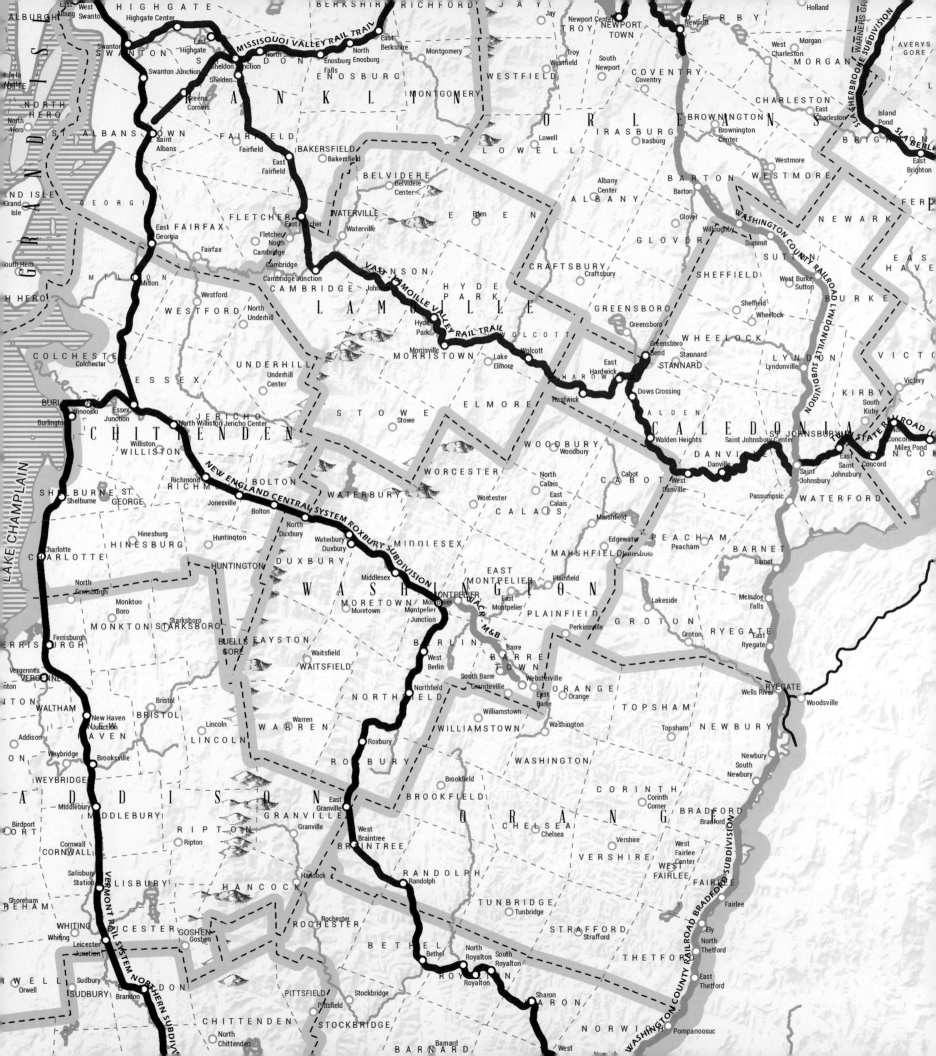

From Takeoff to Landing

Avinor AS

Gardermoen, Oslo, Norway

**By Halvor S. Aasen
and Andreas M. Edvardsen**

Contact

Andreas Melsom Edvardsen,
andreas.edvardsen@avinor.no

Software

ArcGIS Desktop 9.3.1 SP 1 Build 3500, PLTS for
ArcGIS—Foundation 9.3.1 and PLTS for ArcGIS—
Aeronautical Solution 9.3.1, and AGC 3.3.0.3

Data Sources

Avinor, Norwegian Mapping Authority

These maps are used during the different phases of a flight. Each map has connected text that explains more details. The maps are produced by Avinor AS and published in the Aeronautical Information Publication (AIP) of Norway. Avinor is a state-owned limited company that owns and operates forty-six airports in Norway, fourteen in association with the Royal Norwegian Air Force, and is responsible for air traffic control services in Norway. Avinor also operates three area control centers: Bodø Air Traffic Control Center, Stavanger Air Traffic Control Center, and Oslo Air Traffic Control Center.

Courtesy of Avinor AS.

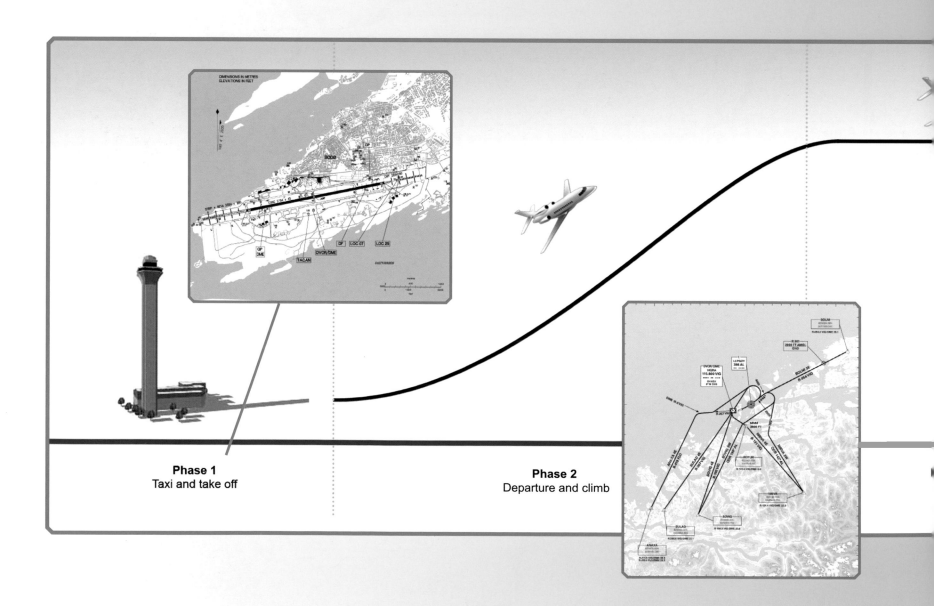

Phase 1
Taxi and take off

Phase 2
Departure and climb

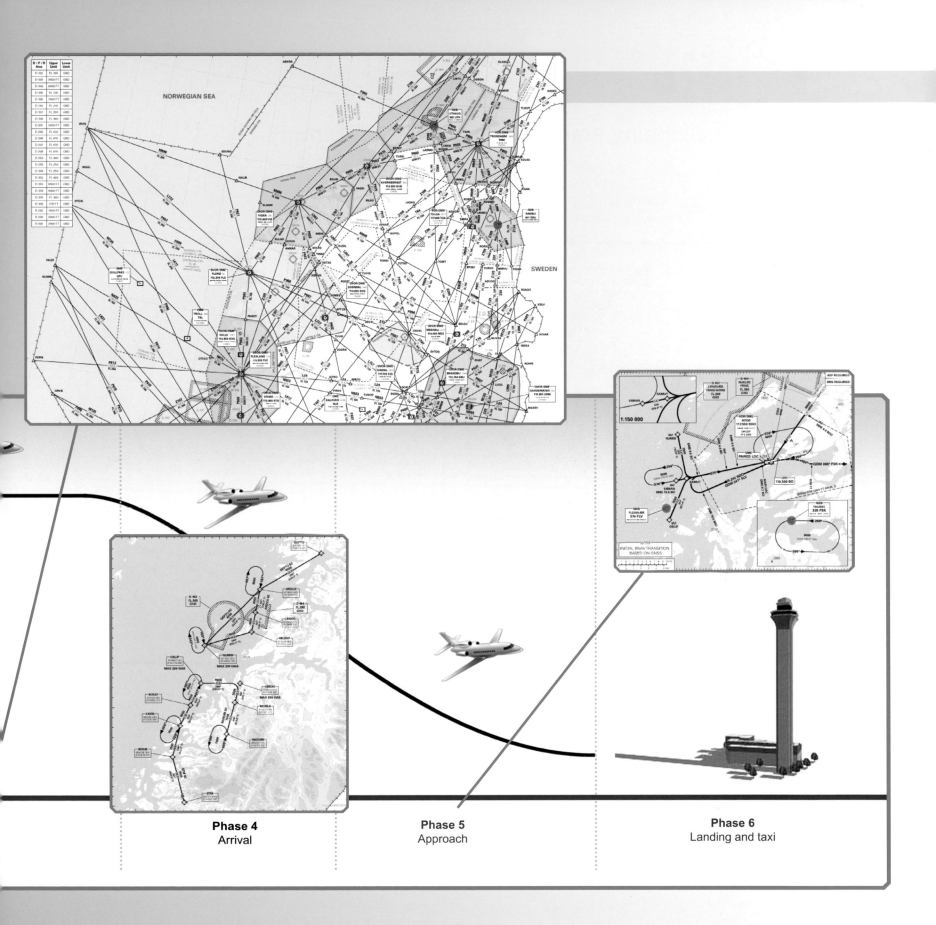

Phase 4
Arrival

Phase 5
Approach

Phase 6
Landing and taxi

New Trans-Arctic Shipping Routes Navigable by Midcentury

UCLA

Los Angeles, California, USA

By Scott Stephenson, Laurence Smith, and Matt Zebrowski

Contact

Scott Stephenson, stephenson@ucla.edu

Software

ArcGIS 10.1 for Desktop, Python 2.7

Data Sources

ACCESS1.0, ACCESS1.3, CCSM4, GFDL-CM3, HADGEM2-CC, IPSL-CM5A-MR, MPI-ESM-MR

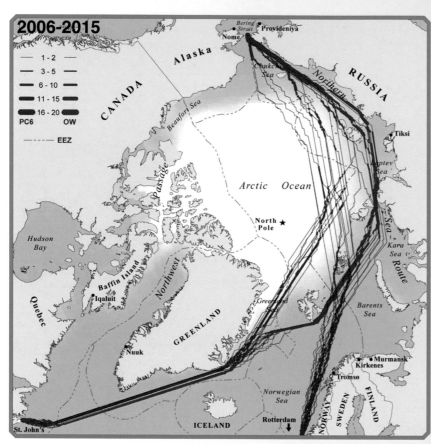

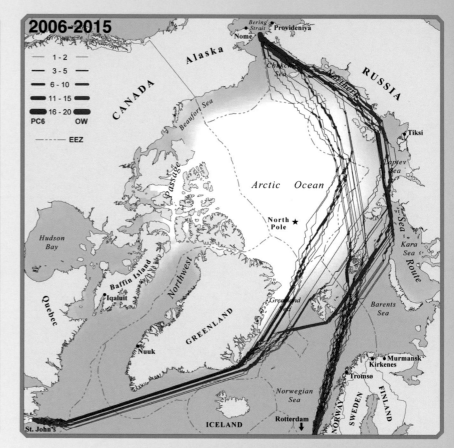

These maps illustrate optimal September navigation routes for hypothetical ships seeking to cross the Arctic Ocean between the North Atlantic (Rotterdam, the Netherlands, and St. John's, Newfoundland) and the Pacific (Bering Strait) during consecutive years 2006–2015 and 2040–2059. The maps reflect global climate model projections of sea ice concentration and thickness assuming representative concentration pathways (RCPs) 4.5 (medium–low radiative forcing) and 8.5 (high radiative forcing) climate change scenarios. The RCPs, used for climate modeling and research, describe several possible climate futures that depend on how much greenhouse gas is emitted in the years to come

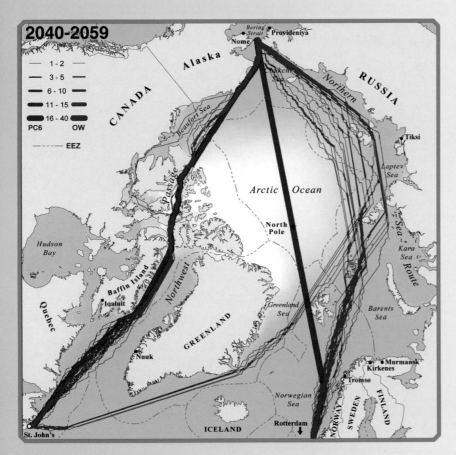

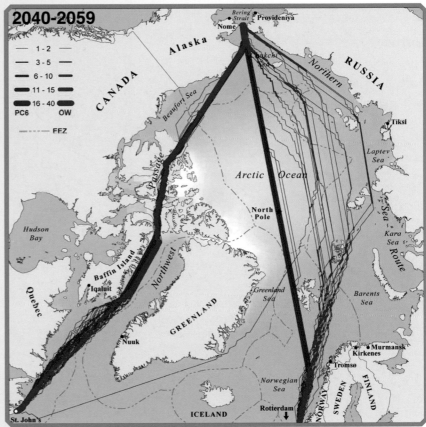

Red lines indicate the fastest available trans-Arctic routes for Polar Class 6 ships; blue lines indicate the fastest available transits for common open water ships. Where overlap occurs, line weights indicate the number of successful transits using the same navigation route. Dashed lines indicate national 200-nm exclusive economic zone (EEZ) boundaries;

white backdrops indicate period-average sea ice concentrations in 2006–2015 and 2040–2059. These maps first appeared in Proceedings of the National Academy of Sciences.

Courtesy of Scott Stephenson, Laurence Smith, and Matt Zebrowski.

Multimodal Site Selection

RS&H Inc.

Jacksonville, Florida, USA

By Patricia Richters

Contact

Patricia Richters, patricia.richters@rsandh.com

Software

ArcGIS 10.1 for Desktop, CommunityViz

Data Sources

St. Johns County, Sunshine Bus Company

The last passenger train stopped in St. Augustine nearly 50 years ago, and finding a way to connect the six million annual visitors to this Old World city and providing a more sustainable transportation option for regional residents continue to be major objectives for Northeast Florida. With the potential for a variety of passenger rail service operations and supporting transit connections in the short and long term, the St. Augustine community wanted a fresh, objective analysis for a center based on measurable demographic, economic, and transportation criteria.

Site options were selected by a GIS-based suitability analysis using weighted criteria. Factors considered included population; employment; accessibility via rail, plane, car, bus, trolley, and bike; and environmental impacts. Government officials, economic development experts, community activists, rail historians, and interested citizens formed a stakeholder group to identify potential sites, weight selection criteria, and evaluate parcels, ensuring a thorough and neutral assessment.

A formal suitability analysis identified potential areas of focus, or "hot spot zones." This spatial-based process determined which locations are best suited for certain uses. ArcGIS planning analysis software allowed each of the stakeholder-developed criteria to be integrated into a grid cell, which functions as a "suitability layer," and weighted relative to its importance in identifying the most suitable location for a successful multimodal center.

Courtesy of Patricia Richters.

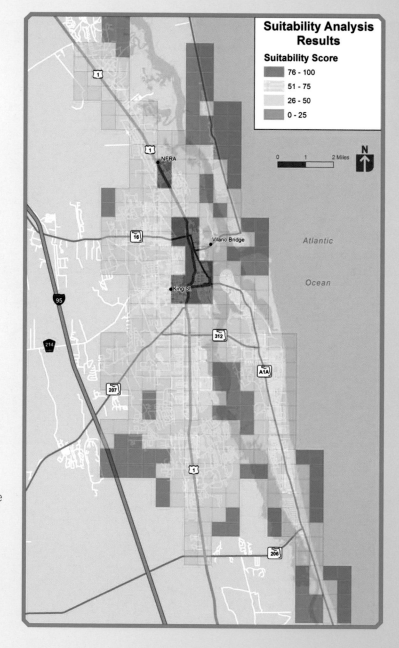

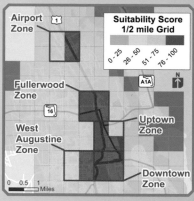

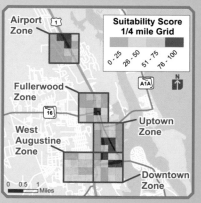

2013 Vermont Transportation Resources Map

Vermont Agency of Transportation

Montpelier, Vermont, USA

By Gary S. Smith

Contact

Johnathan Croft, johnathan.croft@state.vt.us

Software

ArcGIS 10.1 for Desktop

Data Sources

Vermont Agency of Transportation, Vermont Center for Geographic Information, Vermont Agency of Natural Resources, US Geological Survey

This map displays the transportation infrastructure across the State of Vermont and is built on a geographic base that locates prominent places, political boundaries, mountain peaks, and water bodies greater than 25 acres in size. The legend is a custom creation intended to facilitate viewer understanding of the statewide presentation. In addition to providing Vermont state plane coordinate tic marks to help locate features, a simpler x,y coordinate reference grid cell system was built into the map's border. Unlike maps of the past that required the creation of missing data, this map was prepared from data layers created and maintained within the Vermont Agency of Transportation or other parts of state government. While the obvious intent of this map is to illustrate the state's road and rail network, ferry crossings, airports, and other supporting resources, this map also could provide the framework for other thematic presentations, such as recreation.

Courtesy of Vermont Agency of Transportation.

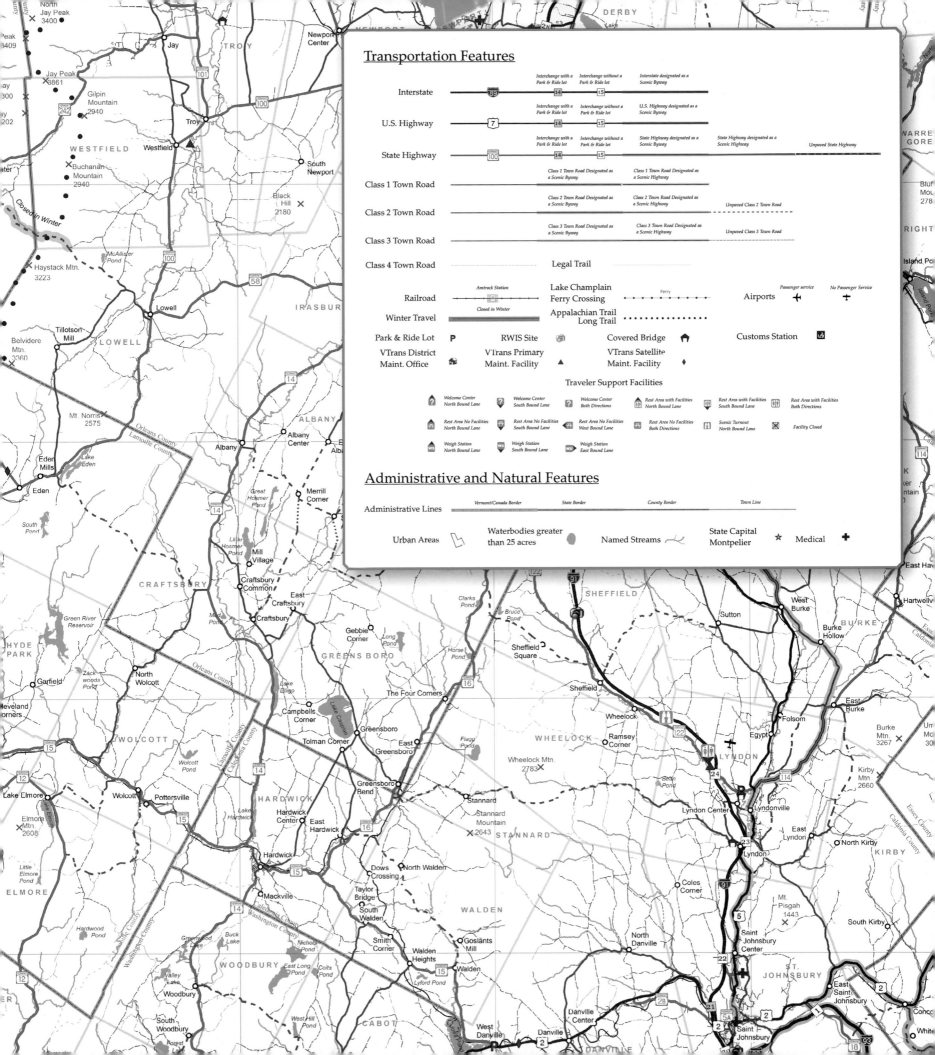

Transportation Features

		Interchange with a Park & Ride lot	Interchange without a Park & Ride lot	Interstate designated as a Scenic Byway
Interstate	89	14	15	

		Interchange with a Park & Ride lot	Interchange without a Park & Ride lot	U.S. Highway designated as a Scenic Byway
U.S. Highway	7	14	15	

		Interchange with a Park & Ride lot	Interchange without a Park & Ride lot	State Highway designated as a Scenic Byway	State Highway designated as a Scenic Highway	Unpaved State Highway
State Highway	100	14	15			

Class 1 Town Road — Class 1 Town Road Designated as a Scenic Byway — Class 1 Town Road Designated as a Scenic Highway

Class 2 Town Road — Class 2 Town Road Designated as a Scenic Byway — Class 2 Town Road Designated as a Scenic Highway — Unpaved Class 2 Town Road

Class 3 Town Road — Class 3 Town Road Designated as a Scenic Byway — Class 3 Town Road Designated as a Scenic Highway — Unpaved Class 3 Town Road

Class 4 Town Road   Legal Trail

Railroad — Amtrack Station

Winter Travel — Closed in Winter

Lake Champlain Ferry Crossing — Ferry

Appalachian Trail / Long Trail

Airports — Passenger service / No Passenger Service

Park & Ride Lot — **P**

RWIS Site

Covered Bridge

Customs Station

VTrans District Maint. Office

VTrans Primary Maint. Facility ▲

VTrans Satellite Maint. Facility ♦

Traveler Support Facilities

Welcome Center North Bound Lane

Welcome Center South Bound Lane

Welcome Center Both Directions

Rest Area with Facilities North Bound Lane

Rest Area with Facilities South Bound Lane

Rest Area with Facilities Both Directions

Rest Area No Facilities North Bound Lane

Rest Area No Facilities South Bound Lane

Rest Area No Facilities West Bound Lane

Rest Area No Facilities Both Directions

Scenic Turnout North Bound Lane

Facility Closed

Weigh Station North Bound Lane

Weigh Station South Bound Lane

Weigh Station East Bound Lane

Administrative and Natural Features

Administrative Lines — Vermont/Canada Border — State Border — County Border — Town Line

Urban Areas

Waterbodies greater than 25 acres

Named Streams

State Capital Montpelier ★

Medical ✚

San Bernardino Critical Routes

San Bernardino County

San Bernardino, California, USA

By Brent Rolf

Contact

Brent Rolf, brolf@isd.sbcounty.gov

Miles Wagner, San Bernardino County Office of Emergency Services,
miles.wagner@oes.sbcounty.gov

Software

ArcGIS 10.1 for Desktop

Data Sources

San Bernardino County, US Geological Survey, local cities

The San Bernardino County Operational Area Critical Route Committee
developed critical routes for first responders and residents to use in
the event of a significant incident or disaster within the county. Local
jurisdictions are able to use the maps to prioritize clearing/repairing
local roadways following the priorities shown on the map during times of
disaster.

Additionally, the committee identified certain features and attributes to
place on the map that will help incident commanders manage an incident,
assign first responders to designated incoming routes still available
after an incident, and advise the general public of evacuation routes if
necessary. Other routes could be designated to the nearest available
evacuation staging areas, hospitals, or points of distribution or other
locations where services would be located.

Critical route maps are used in disaster planning and incident
response throughout San Bernardino County and in some cases will be
used for incidents occurring in neighboring counties that have a local
impact.

Courtesy of Brent Rolf.

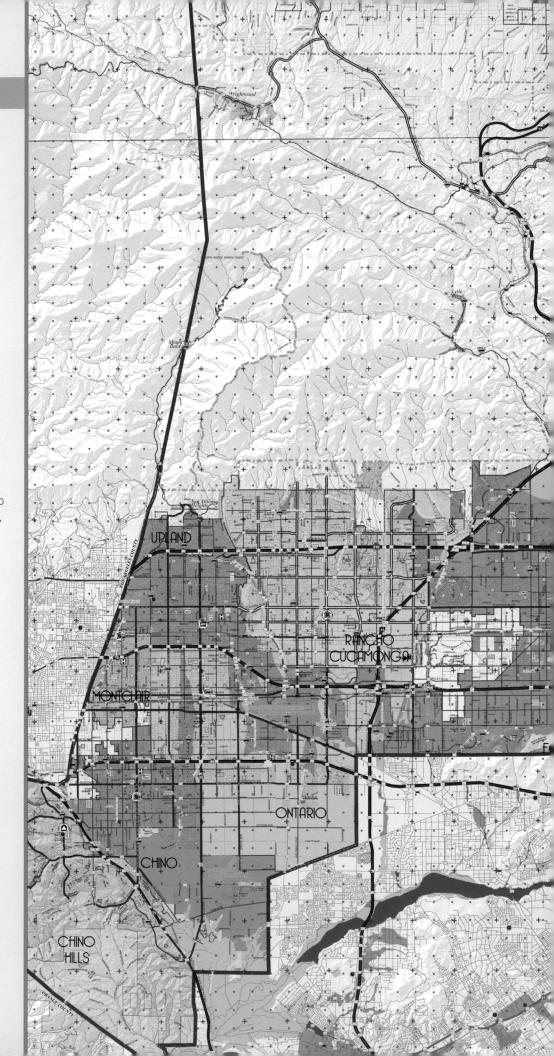

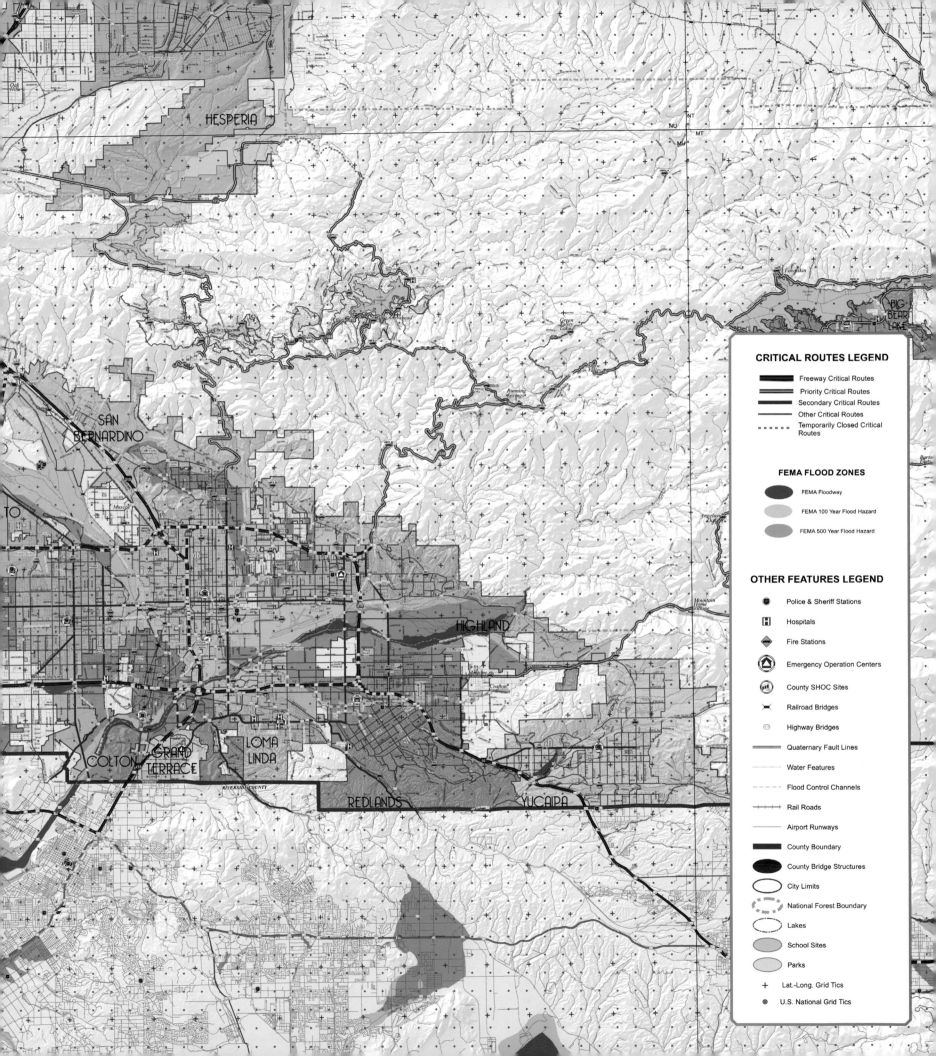

CRITICAL ROUTES LEGEND

▬▬▬ Freeway Critical Routes
▬▬▬ Priority Critical Routes
▬▬▬ Secondary Critical Routes
▬▬▬ Other Critical Routes
▪ ▪ ▪ Temporarily Closed Critical Routes

FEMA FLOOD ZONES

⬤ FEMA Floodway
⬤ FEMA 100 Year Flood Hazard
⬤ FEMA 500 Year Flood Hazard

OTHER FEATURES LEGEND

🔲 Police & Sheriff Stations
Ⓗ Hospitals
◆ Fire Stations
⬢ Emergency Operation Centers
⬡ County SHOC Sites
⋈ Railroad Bridges
▦ Highway Bridges
▬▬ Quaternary Fault Lines
⋯⋯ Water Features
— — — Flood Control Channels
++++++ Rail Roads
▬▬ Airport Runways
▬▬▬ County Boundary
⬤ County Bridge Structures
◯ City Limits
⬭ National Forest Boundary
◯ Lakes
⬤ School Sites
⬤ Parks
+ Lat.-Long. Grid Tics
⊛ U.S. National Grid Tics

HESPERIA

BIG BEAR LAKE

SAN BERNARDINO

TO

HIGHLAND

COLTON
GRAND TERRACE
LOMA LINDA

RIVERSIDE COUNTY

REDLANDS
YUCAIPA

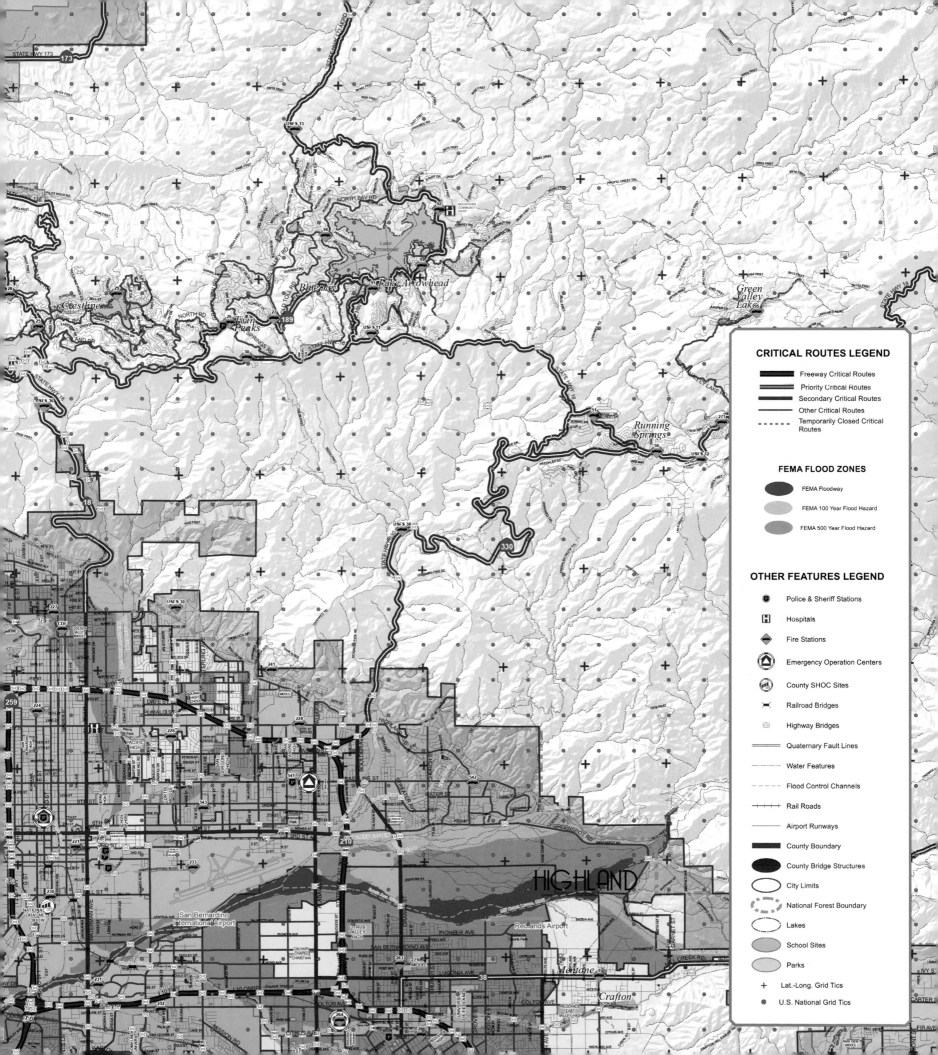

CRITICAL ROUTES LEGEND

Freeway Critical Routes
Priority Critical Routes
Secondary Critical Routes
Other Critical Routes
Temporarily Closed Critical Routes

FEMA FLOOD ZONES

FEMA Floodway
FEMA 100 Year Flood Hazard
FEMA 500 Year Flood Hazard

OTHER FEATURES LEGEND

Police & Sheriff Stations
Hospitals
Fire Stations
Emergency Operation Centers
County SHOC Sites
Railroad Bridges
Highway Bridges
Quaternary Fault Lines
Water Features
Flood Control Channels
Rail Roads
Airport Runways
County Boundary
County Bridge Structures
City Limits
National Forest Boundary
Lakes
School Sites
Parks
Lat.-Long. Grid Tics
U.S. National Grid Tics

ArcGIS for Windows Mobile in Work Force Management System

ČEZ Distribuční služby, s.r.o. (ČEZ Distribution Services Ltd.)

Ostrava, Moravskoslezsky kraj,

Czech Republic

By Pontech s.r.o. and ČEZ Distribucní služby s.r.o.

Contact

Jakub Sigmund, jakub.sigmund@cez.cz

Software

ArcGIS 10.0 for Desktop, ArcGIS 10.0 for Windows Mobile

Data Sources

ČEZ Group, StreetMap for ArcGIS for Windows Mobile NAVTEQ Europe 2012 Release 2

ČEZ Group is the largest electric utility provider in central Europe, serving approximately seven million customers in the Czech Republic, Romania, and Bulgaria. Following the acquisition of three distribution companies in Bulgaria and one in Romania, two Polish power plants, and one Bulgarian power plant, ČEZ Group has become a multinational enterprise comprising over ninety Czech and foreign companies.

ČEZ is implementing GIS as a core enterprise system. ČEZ/Energy-Utilities' ArcGIS/ArcFM system provides editing and viewing access to more than 300 concurrent desktop users, 800 mobile users, and 400 concurrent users accessing web applications. The system supports engineering, operations, compliance, work force management, and other activities.

In 2012, ČEZ Distribution Services, a member of the ČEZ Group, implemented a mobile GIS application based on ArcGIS 10.0 for Windows Mobile. The application helps the field workers orient themselves in the terrain and locate the task in the power supply and reduces the need to produce paper maps that constantly had to be updated, reprinted, saving both cost and time.

Courtesy of Jakub Sigmund.

Dalkia Implements ArcGIS App for Smartphones

Dalkia Czech Republic
Ostrava, Czech Republic

By Stanislav Splichal

Contact

Stanislav Splichal, stanislav.splichal@dalkia.cz

Software

ArcGIS 10.0 for Desktop, ArcGIS Apps for Smartphones

Data Source

Dalkia Czech Republic

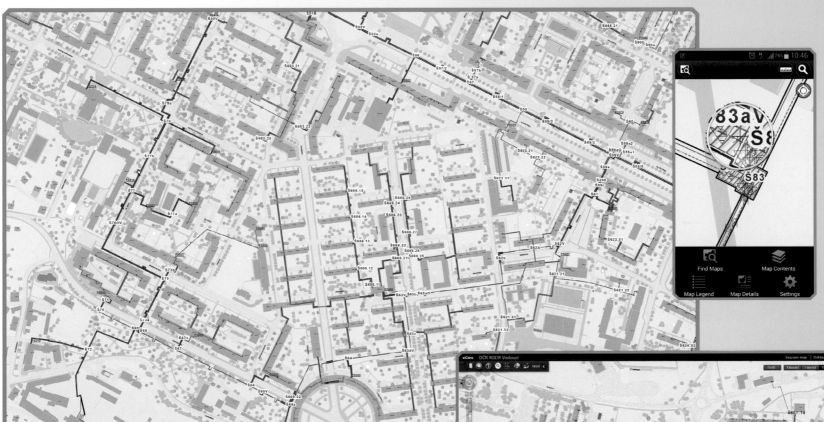

Dalkia is the biggest independent cogeneration producer of heat and power in the Czech Republic. It manages more than 1,850 kilometers of primary and secondary networks that involve constant monitoring. To extend the enterprise use of GIS and help the field workers, Dalkia decided to implement mobile GIS using mobile devices with Android OS.

ArcGIS for Android, a free, downloadable mobile application, uses data published by ArcGIS for Server. The users can choose from several map projects containing geographic and attribute data. This map shows Dalkia's heating distribution network in Ostrava city. On the network, there are several types of installations (substations, valve chambers) on the map as well as customer supply points. The basemap for this map composition is called "Digital Technical Map" of Ostrava city. The map composition is accessible not only for mobile workers in the field but also for other Dalkia employees through the Flex Viewer web mapping application.

The mobile solution helps the field workers quickly identify facilities in the terrain, easily mark out the distribution network, speed up repairs during emergency situations, and easily orient themselves in the field. In addition, workers can display attached contracts, schemes, documents, photographs, and plans according to the user's role.

Courtesy of Dalkia Czech Republic.

Culver City's Storm Drain Map Book

City of Culver City
Culver City, California, USA

By Marcos Mendez

Contact

Johnnie Griffing, John.Griffing@culvercity.org

Software

ArcGIS 10.1 for Desktop

Data Source

City of Culver City GIS data

Culver City's Storm Drain Map Book shows manholes, catch basins, and underground main and lateral lines. The city, being completely surrounded by other jurisdictions, needed to identify ownership for storm drain features. Once the majority of the storm drain infrastructure had been developed in GIS, map book creation began using Data Driven Pages. Map book pages were delineated using a grid provided by the surrounding City of Los Angeles Public Works Department. The page number scheme was modified for Culver City's use, and adopting the same grid as Los Angeles facilitates comparing map data between the two cities. Each page was designed to contain as much information as possible and organized for ease of use by engineers and field crews. An annotation layer was developed to ensure legibility within the limitations of the page size. The printed books are laminated and bound for rugged field use.

Courtesy of City of Culver City.

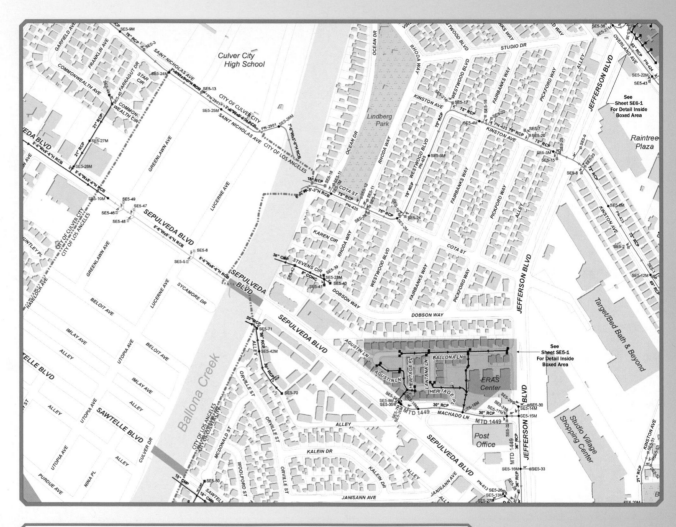

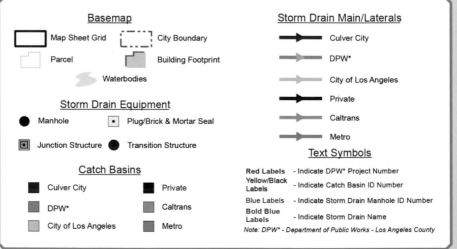

Dutch Drinking Water Hardness from 1982 to 2012

KWR Watercycle Research Institute
Nieuwegein, Utrecht, Netherlands

By Bernard Raterman and Bas Hofs

Contact
Bernard Raterman, bernard.raterman@ kwrwater.nl

Software
ArcGIS 10.0 for Desktop

Data Sources
Dutch water companies, Vewin

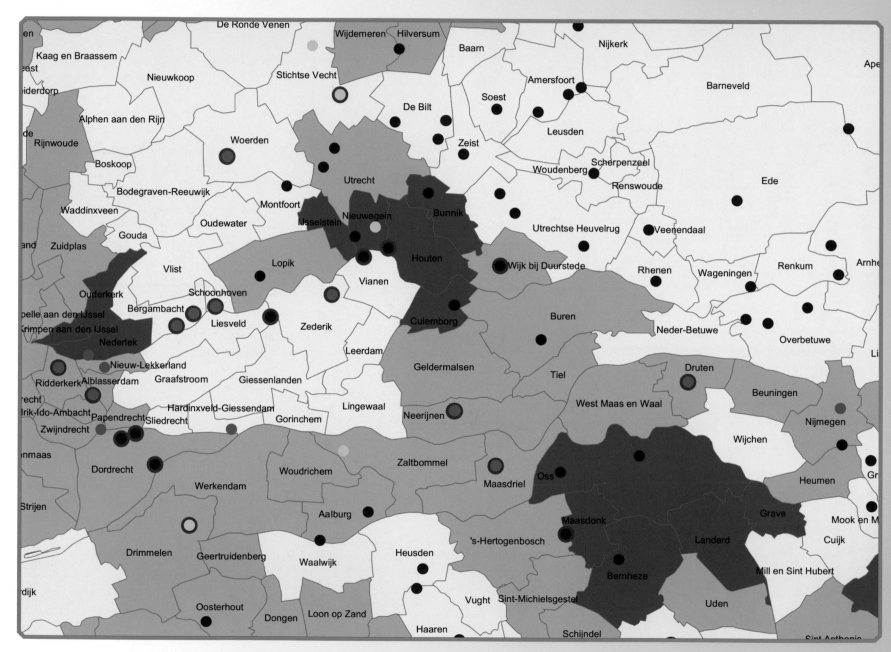

Water hardness is mainly determined by the concentrations of calcium and magnesium. If the aquifer or subsoil around a groundwater abstraction contains a lot of dissolved calcium and magnesium minerals, then the pumped water is naturally hard. In the Netherlands, water hardness is usually expressed in so-called German degrees (° DH). In the late 1960s, there was an increasing demand for softened water, and after 1984, there was a sharp increase in the use of central water softening in the Netherlands. Softening is now used because it leads to less lime scaling and therefore less energy consumption in hot water systems at home. Other benefits are a reduction in the concentrations of lead and copper in drinking water and a reduction of the required amount of detergent and soap.

Courtesy of KWR Watercycle Research Institute.

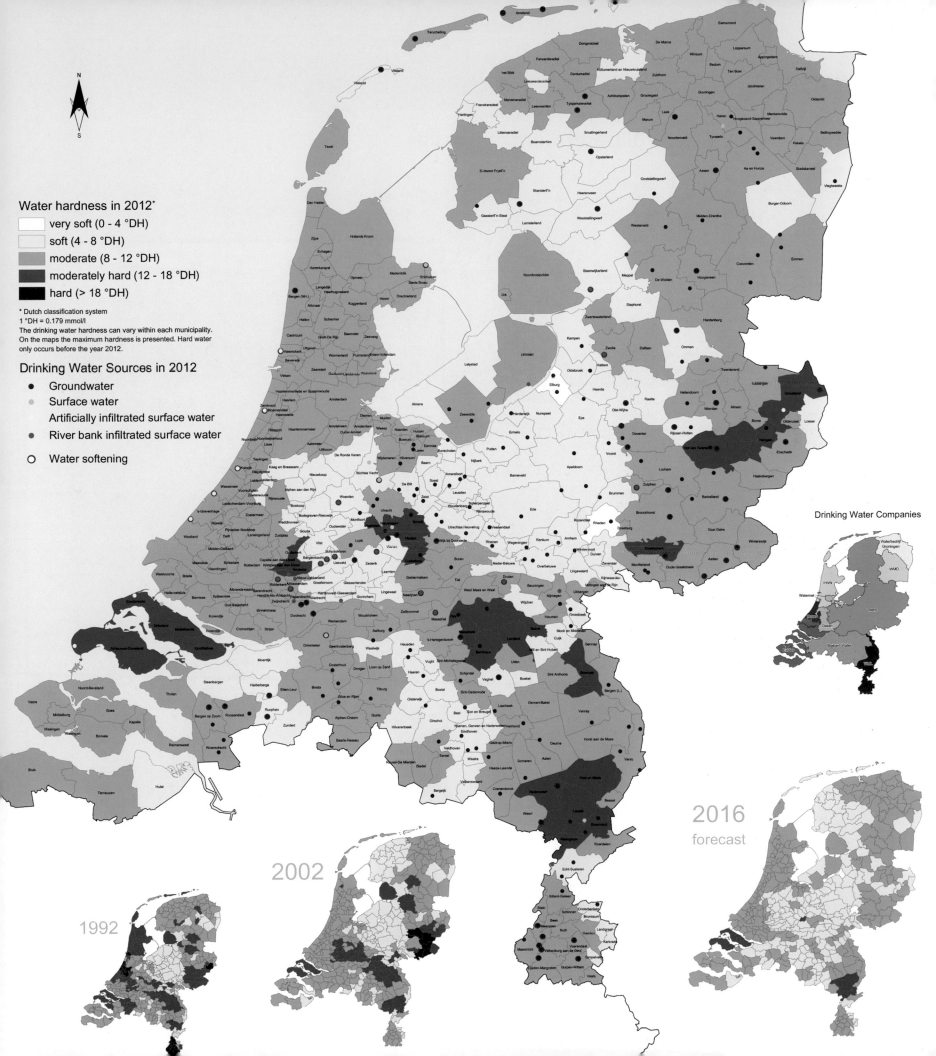

Water hardness in 2012*

- very soft (0 - 4 °DH)
- soft (4 - 8 °DH)
- moderate (8 - 12 °DH)
- moderately hard (12 - 18 °DH)
- hard (> 18 °DH)

* Dutch classification system
1 °DH = 0.179 mmol/l
The drinking water hardness can vary within each municipality.
On the maps the maximum hardness is presented. Hard water
only occurs before the year 2012.

Drinking Water Sources in 2012

- Groundwater
- Surface water
- Artificially infiltrated surface water
- River bank infiltrated surface water
- Water softening

Drinking Water Companies

1992

2002

2016
forecast

City of Greeley Water Pollution Control Facility

City of Greeley

Greeley, Colorado, USA

By Scott Cox

Contact

Scott Cox, Scott.Cox@greeleygov.com

Software

ArcGIS 10.0 for Desktop

Data Source

City of Greeley

In 1904, the City of Greeley obtained senior water rights 35 miles upstream from town on the Cache la Poudre River. The land that accompanied the water rights became the site of Greeley's Bellvue water treatment plant. Today, a series of four pipelines (one of which is 60 inches in diameter) are used to bring drinking water from high on the river, down to the city. Greeley's Water Pollution Control Facility is at the other end of that system. After drinking water is distributed to residences and businesses, it is collected by the sanitary sewer system and flows to the facility to be treated before being reintroduced to the Cache la Poudre River.

This map shows many of the systems required to run the facility, including storm water runoff, flood mitigation, biosolids extraction, and, of course, water treatment. An average of 8 to 9 million gallons of wastewater from homes, businesses, and industries within the City of Greeley are treated every day at the Water Pollution Control Facility.

Courtesy of City of Greeley, Colorado 2013.

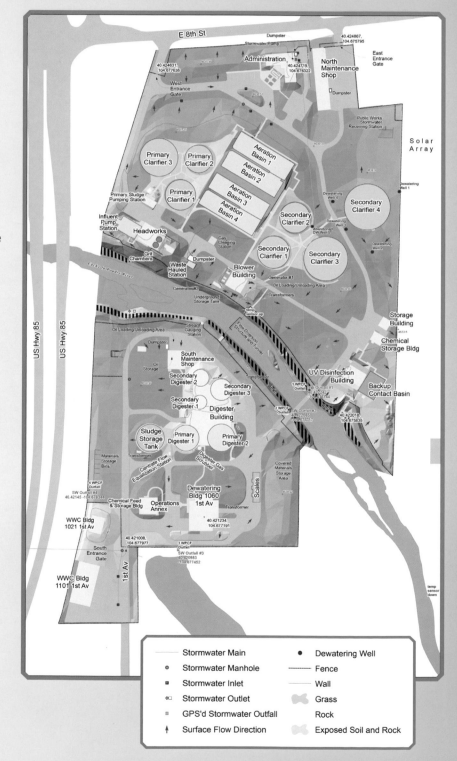

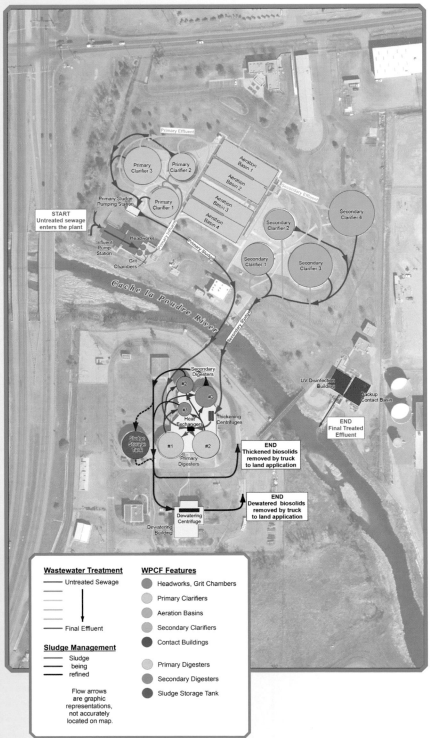

Wasterwater Treatment

Untreated Sewage

Final Effluent

Sludge Management

Sludge
being
refined

Flow arrows
are graphic
representations,
not accurately
located on map.

WPCF Features

Headworks, Grit Chambers

Primary Clarifiers

Aeration Basins

Secondary Clarifiers

Contact Buildings

Primary Digesters

Secondary Digesters

Sludge Storage Tank

Oro Valley Water Utility Basemaps

Oro Valley Water Utility

Oro Valley, Arizona, USA

By Irene Swanson

Contact

Irene Swanson, iswanson@orovalleyaz.gov

Software

ArcGIS for Desktop

Data Source

Oro Valley Water Utility, Pima County

The Town of Oro Valley is located approximately 6 miles north of Tucson, Arizona, and is situated in the northeastern corner of Pima County. The Oro Valley Water Utility is municipally owned and has been in operation since 1996. The utility has two systems, the central system, which lies inside the town limits, and the countryside water system west of the town. The total water service area is about 33.4 square miles.

Until 2002, the only maps available to the field and office staff were construction drawings. A consultant was hired to convert the drawings to an AutoCAD mapping system, and shortly after that, the GIS conversion project started. With one GIS analyst on staff, the project was completed and in production mode within three years.

Based on Esri's enterprise geodatabase environment, the utility currently has a complete and accurate water basemap atlas of about 100 maps. Each map covers one half of a section (0.5 mile) showing the water distribution system as well as additional basemap information.

Courtesy of Oro Valley Water Utility.

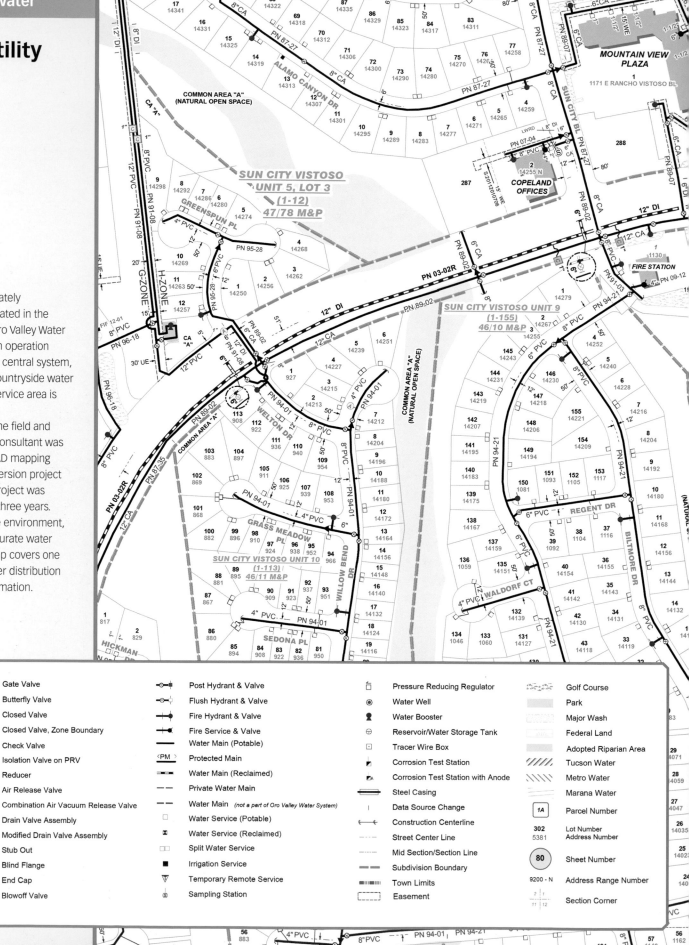

⊗	Gate Valve		Post Hydrant & Valve		Pressure Reducing Regulator		Golf Course
⊗	Butterfly Valve		Flush Hydrant & Valve		Water Well		Park
⊖	Closed Valve		Fire Hydrant & Valve		Water Booster		Major Wash
	Closed Valve, Zone Boundary		Fire Service & Valve		Reservoir/Water Storage Tank		Federal Land
	Check Valve		Water Main (Potable)		Tracer Wire Box		Adopted Riparian Area
	Isolation Valve on PRV	‹PM›	Protected Main		Corrosion Test Station		Tucson Water
▶	Reducer		Water Main (Reclaimed)		Corrosion Test Station with Anode		Metro Water
	Air Release Valve		Private Water Main		Steel Casing		Marana Water
	Combination Air Vacuum Release Valve		Water Main *(not a part of Oro Valley Water System)*		Data Source Change	1A	Parcel Number
	Drain Valve Assembly	□	Water Service (Potable)		Construction Centerline	302 / 5381	Lot Number / Address Number
	Modified Drain Valve Assembly		Water Service (Reclaimed)		Street Center Line		
	Stub Out	□□	Split Water Service		Mid Section/Section Line	80	Sheet Number
	Blind Flange	■	Irrigation Service		Subdivision Boundary		
	End Cap	▽	Temporary Remote Service		Town Limits	9200 - N	Address Range Number
	Blowoff Valve		Sampling Station		Easement		Section Corner